Models for Embryonic Periodicity

Monographs in Developmental Biology

Vol. 24

Series Editor
H.W. Sauer, College Station, Tex.

Basel · Freiburg · Paris · London · New York · New Delhi · Bangkok · Singapore · Tokyo · Sydney

Models for Embryonic Periodicity

Lewis I. Held, Jr.
Department of Biological Sciences, Texas Tech University, Lubbock, Tex.

11 figures and 2 tables, 1992; 2nd printing, 1994

KARGER

Basel · Freiburg · Paris · London · New York · New Delhi · Bangkok · Singapore · Tokyo · Sydney

Monographs in Developmental Biology

Founded 1969 by A. Wolsky, New York, N.Y.

Vol. 24: 2nd printing, 1994

Library of Congress Cataloging-in-Publication Data
Held, Lewis I., 1951 –
Models for embryonic periodicity / Lewis I. Held, Jr.
p. cm. – (Monographs in developmental biology; vol. 24)
Includes bibliographical references.
1. Developmental biology. 2. Embryonic periodicity. 3. Pattern formation (Biology).
4. Developmental cytology. I. Title.
II. Series: Monographs in developmental biology: v. 24.
QH491.H45 1992   92–13644
574.3′01′1 – dc20   CIP
ISBN 3–8055–6008–7

Bibliographic Indices
This publication is listed in bibliographic services, including Current Contents® and Index Medicus.

Drug Dosage
The authors and the publisher have exerted every effort to ensure that drug selection and dosage set forth in this text are in accord with current recommendations and practice at the time of publication. However, in view of ongoing research, changes in government regulations, and the constant flow of information relating to drug therapy and drug reactions, the reader is urged to check the package insert for each drug for any change in indications and dosage and for added warnings and precautions. This is particularly important when the recommended agent is a new and/or infrequently employed drug.

Contents

Preface

Lord Kelvin once remarked that he never fully understood a process until he could make a mechanical model of it [431]. This is a book full of models, some of which have profoundly influenced the history of developmental biology. The particular themes of the book, its iconoclastic style, and its focus on cybernetics are attributable to my own academic odyssey. My first exposure to model-building was when I worked as a computer programmer (under Seymour Papert and Marvin Minsky) in the Artificial Intelligence Laboratory at MIT where I was an undergraduate. The lab group was interested in how humans think, and most of the members designed programs to enable computers to 'converse', 'see', play chess, etc. Others, including myself, wrote interactive programs for teaching children about scientific principles. I created a simulated microworld of reptilian evolution where the user would select an environment, and the reptile population would attempt to adapt by randomly changing the ranges of variation of its anatomical parameters (via 'mutations'). What amazed me was how easy it was to reduce seemingly complicated structures (e.g. feathers) to simple equations that could be drawn by graphic algorithms. Only later did I encounter D'Arcy Thompson's 'On Growth and Form' [882] which makes this same point for anatomical shapes in general.

Had it not been for an apprenticeship with David Botstein and Ira Herskowitz, where I first encountered the joys (and sorrows) of biological research (my project concerned T7 phage genetics), I might never have left computer science. After applying to and being accepted in the graduate program in UC Berkeley's Molecular Biology Department, I was faced with the choice of an advisor and a project. One day, in our core course, John Gerhart delivered a fascinating guest lecture on Escherian symmetries in virus heads, and soon thereafter I asked to work in his lab. At the time, he was investigating both frog and fly development, and I chose to work on flies because of their intriguing patterns of bristles. Chiyo Tokunaga, an expert geneticist, had recently joined the lab after collaborating for many years with Curt Stern, a pioneer in the field of developmental genetics. She would tell me stories of the early days in Stern's lab, when he was formulating his Prepattern Hypothesis. His theory was rapidly being eclipsed by Lewis Wolpert's Positional Information Hypothesis, and it was illuminating to debate the pros and cons of the two paradigms with Chiyo and John. One of the essay questions in John's developmental biology course (co-taught by Gunther Stent) asked about a corollary of Wolpert's 'French Flag Problem': how could a concentration gradient of a chemical cause cells to produce blue, white, or red pigments in different zones? My answer was that if cells had vesicles containing pigments of different colors, then the chemical could cause osmotic swelling, and different colors could be released due to different bursting thresholds. I never liked that answer (though I did get an 'A' on the test) nor any of the then-popular explanations for how cells interpret positional information. I finished my dissertation in 1977, still skeptical about the ability of the new theory to explain patterns containing large numbers of identical elements such as bristles.

My postdoctoral years were spent in the think tank where the Polar Coordinate Model was born: the Developmental Biology Center at UC Irvine. The Center was a cauldron of ideas concocted by the faculty (including my sponsors Peter Bryant and Howard Schneiderman), the postdocs and graduate students, ... and repeatedly stirred and spiced by a parade of visiting scientists (cf. the book 'Cellular Basis of Morphogenesis' [223] to get a feeling for the ongoing ferment in this field). To make sense of the panoply of different 'patterning' theories, I began classifying them using a framework that I had devised as a graduate student. This book is the culmination of that effort, tempered by years of trying to teach squirming undergraduates about the wonders of gradients and clockfaces. I offer it as a field guide for others who have also felt lost in the wild menagerie of strange models that have lately seemed to multiply without limit.

Constructive comments on the manuscript were kindly furnished by Larry Blanton, Richard Campbell, John Gerhart, Kent Rylander, and Helmut Sauer (series editor). Technical jargon has been avoided wherever possible, but a college-level understanding of embryological principles is essential. Newcomers to developmental biology may find the book's numerous citations useful as entry points to the field's vast literature. Finally, although I have endeavored to achieve an ecumenical scope, my parochial background as a fly researcher colors the text in many places. Do not interpret this slant as bias. The next breakthrough in this field could as easily come from a creature with green leaves as one with six legs.

Lubbock, Tex., December, 1991   *Lewis I. Held, Jr.*

Introduction

Periodicity

Fertilized eggs bear little resemblance to the multicellular adults that they become. Aside from their smaller size, eggs are typically ovoid and featureless whereas adults have complex shapes and anatomies. Most notably, eggs are single cells while adults contain tens or hundreds of different cell types (nerve cells, muscle cells, etc.) arranged in intricate patterns. The process whereby one cell generates many types of cells is called 'differentiation' [948] (literally the acquisition of differences among cells), and the spatial control of differentiation is termed 'pattern formation' [927]. The question of how patterns originate is the Gordian Knot of developmental biology.

Anatomical patterns may have either unique or 'repeated' elements. For instance, each half of your face has 4 unique elements: an eye, an ear, and half of a nose and mouth. In contrast, your hand has 5 similar digits, 2 or 3 phalanges per digit, and dozens of evenly spaced ridges on each fingertip. Repeated elements that are arranged at regular intervals constitute a 'periodic' pattern [30, 136, 310, 573, 706], and such patterns are extremely common in animals and plants, e.g. teeth and ribs, zebra stripes and leopard spots, cat whiskers and dog teats, fish scales and bird feathers, tree branches and flower petals, caterpillar segments and butterfly wing veins. Because 'periodicity' [305, 949] is such an important organizing principle in anatomy, it is worthy of study in its own right. Surprisingly however, there has not been a comprehensive treatise on this topic since William Bateson's classic 1894 monograph 'Materials for the Study of Variation' [46] nearly 100 years ago.

Many developing organisms can produce constant patterns despite the surgical addition or removal of tissue. The theoretical challenge posed by this 'regulative' ability has been abstractly formulated as the 'French Flag Problem' [14, 996]: *how can an array of cells generate three different (blue, white, and red) zones of equal width, regardless of the total number of cells?*

A clever idea that solves this riddle was proposed by Hans Driesch [186, 1011] in 1901, and in 1969 Lewis Wolpert [997] formalized the concept as the 'positional information' hypothesis: (1) embryonic cells differentiate based upon information that they receive about their positions within a coordinate system, and (2) the boundary coordinates are specified independently of physical dimensions so that the entire system adjusts to the size of the field that it spans (i.e. the embryo or a part thereof). The details of how the mechanism works will be discussed later. What is important here is that the school of thought that has evolved from the concept of positional information is rooted in – and constrained by – the property of scaling invariance.

An alternative starting point for theorizing is the property of periodicity. Consider the following 'American Flag Problem' (fig. 1): *how can an array of cells generate periodic patterns of elements like the stars and stripes of the American Flag?* Positional information can also solve this riddle, but so can a host of other theoretical mechanisms that will be discussed in this book. A subordinate problem is how some periodic patterns routinely develop a precise number of elements (e.g. 10 fingers, 24 ribs, 32 teeth) or, in terms of the American Flag: *how can an array of cells generate exactly 13 stripes and 50 stars?* Numerical constancy is related to size independence and will be treated as a separate issue (chap. 7).

Types of Patterning Mechanisms

Embryonic cells can be committed to different pathways of development without manifesting any tissue-specific differences. For example, the wings and legs of fruitflies develop from separate groups of cells inside the larva. Prior to metamorphosis, wing and leg cells are indistinguishable cytologically, but when transplanted into the abdominal cavity of a host larva they form only wing or leg structures, respectively. They are said to be 'determined' [408] for different fates and to possess distinct 'states of determination' [339].

Given the notion of 'states' [795, 924, 941, 948] the problem of pattern formation can be reduced to simple mathematical terms. If 'p' designates the position of a cell and 's' is its state of determination or differentiation, then any pattern can be represented as a set of ordered pairs (p, s). For example, if the positions along a line are numbered from 1 to 6, then an alternating pattern of black (B) and white (W) cells can be symbolized as

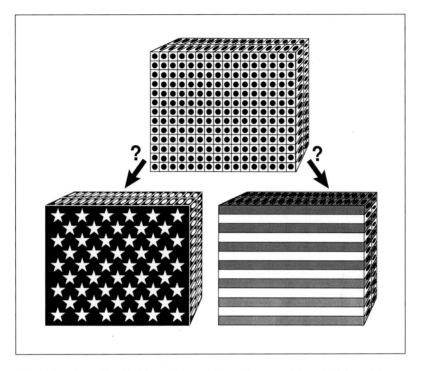

Fig. 1. American Flag Problem. The question of how spatial periodicity originates during development can be abstractly formulated as the 'American Flag Problem': what kinds of mechanisms can allow an array of cells (top) to produce regular patterns – such as a hexagonal (or square) lattice (left) or a set of alternating stripes (right) – where the pattern elements are arranged at uniform intervals? Unlike Wolpert's French Flag Problem [996], an ability to 'regulate' (i.e. restore the entire pattern if part of it is removed) is not demanded, so the set of possible mechanisms is relatively less constrained. A tangential issue (the 'Counting Problem' [559]) is how an exact number of elements (50 stars or 13 stripes) can be reliably generated during development.

'[(1, B), (2, W), (3, B), (4, W), (5, B), (6, W)]'. The general problem thus becomes: 'What causes the correlation of particular values of p and s?' Whenever two entities are correlated in nature (e.g. fire and smoke, thunder and lightning, winter and spring), either one causes the other or both are caused by a third force. For p and s, the possibilities are:

(1) $p \rightarrow s$: The position of a cell causes it to adopt a particular state.

(2) $s \rightarrow p$: The state of a cell causes it to adopt a particular position.

(3) $x \rightarrow p \& s$: Some third agent ('x') causes the correlation of positions

and states. An example of x is cell lineage, since a mother cell can divide directionally (causing the p of each daughter) and bestow instructions (s) asymmetrically.

These causal relationships define distinct classes of mechanisms, and most published models can be assigned unambiguously to a single class (table 1). (This same taxonomic scheme has been advocated by Steinberg and Poole [822].) By contrast, *actual* developmental pathways typically employ *multiple* strategies [180, 369, 459, 541, 646, 967] (see chap. 7). Operationally, the type of mechanism that is used at a given time and place should be discernable by transplanting a cell from one position to another. If the cell changes its fate (adopting a fate appropriate for its new position), then its position was causing its state. If the cell moves back to its original position, then its state was causing its position. If the cell neither changes

Table 1. Differences among pattern formation models, with regard to the creation of spatially periodic patterns

Model or category	Derivative or related models	Distinguishing features
Position-dependent $(p \rightarrow s)$ *class*		The position of each cell (relative to field boundaries or neighboring cells) dictates its state of differentiation
Positional information subclass		Cells know where they are via coordinates which they 'interpret' as particular states of differentiation The coordinate systems allow the patterns to 'regulate'
Gradient Model	Source-Sink Model [24, 149, 996, 997] Double-Gradient Model [14, 156, 157, 395, 744, 997, 998] Phase-Shift Model (wavelength = field length) [124, 144, 310] Gierer-Meinhardt Model (wavelength = field length) [279]	The coordinate system is established (independently of growth) by a scalar variable with fixed boundary values
Polar Coordinate Model [81, 245]	Cartesian-Coordinate intercalation models [153, 441, 445, 448, 747, 977] Discrete-Territory intercalation models [575, 785, 793] (including the Four-Color Wheel Model [311]) Coordinate-free intercalation models [509, 597, 977, 978]	A 'Shortest Route Rule' or 'Smoothing Rule' fills in missing coordinates by intercalary growth
Progress Zone Model [867, 868]	Progress-Zone/Oscillator Model [1011]	Coordinates are assigned temporally as cells exit a growth zone

Table 1 (continued)

Model or category	Derivative or related models	Distinguishing features
Prepattern subclass		Cells assume particular states of determination due to mechanical or chemical signals within the cell layer or by induction from an adjacent cell layer Identical signals are used for identical elements Patterns do not regulate (unless ad hoc assumptions are added)
Physical Force models	Traction Model [36, 351, 662] Periodic Buckling Model [37, 38, 495, 835, 959] Physicochemical Model [288]	Deformations arise at periodic intervals within a tissue layer, causing cells to adopt particular states of determination above a certain threshold of stress or strain
Reaction-Diffusion models	Turing Model [31, 219, 907] Gierer-Meinhardt models (wavelength < 0.5 field length) [279, 570, 571, 573, 579, 580]	Chemicals which have different diffusion rates react, causing an initially uniform chemical distribution to peak at 'wavelength' intervals Above a certain threshold concentration, cells adopt a particular state of determination
Induction Model	Template Model [105, 106, 433, 434]	Periodically arranged cells in one layer induce states of determination in the cells of an apposed layer
Determination wave subclass		States of determination are specified within a zone that traverses an array of cells
Chemical Wave models	Belousov-Zhabotinsky Reaction [227, 470, 619, 643, 976, 979, 980] Liesegang Reaction [364, 373, 611, 919]	Traveling (or standing) waves in the concentration of a diffusible molecule (or precipitate) arise through chemical reactions
Sequential Induction Model	*Dictyostelium* cAMP-Signal Model [171, 172, 600, 635, 707, 729]	Each cell induces a neighboring cell to adopt a particular state
Clock and Wavefront Model [134, 146, 1020]	Pendulum-Escapement Model [136, 574] Progress-Zone/Oscillator Model [1011] Phase-Shift Model (wavelength < 0.5 field length) [124, 144, 310]	Cells oscillate between two states and cease oscillating when a wavefront reaches them
Inhibitory Field and Competence Wave Model	Claxton's Bristle-Spacing Model [116] Ede's Feather-Lattice Model [196] Osborn's Clone Model (teeth) [659] Phyllotaxis models [593, 721, 883] Specific Inhibitor Model [730–732] Specific Activator Model [952] Serial Diversion Model [321] Lateral-Activation/Local-Exclusion Model [573, 580]	Cells are not 'competent' to differentiate before a wavefront reaches them Cells that can adopt a 'preferred' state do so and inhibit neighboring cells from doing so

(Table 1 continued next page.)

Table 1 (continued)

Model or category	Derivative or related models	Distinguishing features
Darwinism subclass		Each cell adopts a state (perhaps randomly) and then examines the states of its neighbors; if it matches a neighbor, then it takes action to correct this 'error'
Cell death models	Edelman's Neural Darwinism Model [202]	Homotypic matches are eliminated by having one of the matched cells die
State change models	Edelman's Topobiology Model [203, 271] Kauffman's Adaptive Antichaos Model [447]	Homotypic matches are eliminated by having one of the matched cells change its state
Rearrangement (s→p) class		Each cell adopts a state (perhaps randomly); the states then cause cells to move until they reach particular locations
Adhesion models	Sperry's Chemoaffinity Model [808] Labeled Pathways Model (insect CNS) [302] Adhesive Hierarchy Model (insect PNS) [57, 300] Synthetic Model (retinotectal projection) [236, 237, 239] Differential Adhesiveness Gradient Model [632] Steinberg's Differential Adhesion Model [820]	The final location of a cell is determined by its ability to adhere to a target cell(s)
Repulsion models	Twitty's Mutual Repulsion Model (chromatophores) [908, 911, 912]	Each cell moves as far as possible away from cells of its own kind
Interdigitation models	Süffert's Interdigitation Model (butterfly scale cells) [862, 1014]	Stripes of unlike cells interdigitate
Chemotaxis models	Snake-Striping Model [622]	Dispersed cells aggregate by mutual attraction
Cell-lineage (x→p,s) class		Cells divide asymmetrically (according to rigid pedigree rules), placing each daughter in a definite position and assigning it a particular state
Quantal Mitosis Model	Cassette Model (yeast mating type) [375]	All cells undergo an asymmetric and polarized 'quantal' mitosis, which assigns left daughters one state and right daughters another
Stem Cell Model	Flip-Flop Feedback Model (leech) [59] Osborn's Clone Model (teeth) [659] Progress-Zone/Oscillator Model [1011] L-System models [518–520, 522]	A cell cyclically changes its state as it divides, causing the states of its daughters to alternate in space as it oscillates in time
Cortical Inheritance Model	Directed Assembly Model [329]	A periodic pattern of molecules is created in the cortical layer of a cell, and each daughter differentiates according to the molecules it inherits

its fate nor moves back, then the third type of mechanism would be indicated, though this result would also be expected after a $p \rightarrow s$ or $s \rightarrow p$ mechanism has been completed.

Vignettes of models from the first category are presented in chapters 1–4. Chapters 5 and 6 discuss models from the second and third categories, respectively. To facilitate comparison, the same linear array of six alternately black or white cells is used in all of the model illustrations, and each model's rules for cellular decisions are listed. Extrapolation to 2 or 3 dimensions is usually left to the reader, as is generalization to patterns where the elements and intervals are multicellular, rather than unicellular. In chapter 7, an attempt is made to distill from the foregoing models a cybernetic 'deep structure' [111, 590] that directs development and constrains evolution. Though far from encyclopedic (for other compendia cf. Meinhardt [573], Murray [619], and Ransom [706]), the sample of models discussed in this book will acquaint the reader with the *diversity* of possible explanations for patterning phenomena. If there is a single lesson to be gleaned from this survey, it is 'Wolpert's Maxim': do not infer process from pattern, since so many processes can produce the same pattern [998]. An awareness of theoretical alternatives can also serve as an antidote for the tendency to shoehorn new data into ill-fitting old paradigms, including the currently reigning theory, which was sired by Wolpert himself.

Chapter 1:
Positional Information Mechanisms

Positional information models belong to the $p \rightarrow s$ category because they postulate a position-dependent assignment of differentiated states. Cells are supposedly informed of their positions, and this information causes them to select particular states according to predetermined rules. Individual models differ only in how they specify positional information.

The Gradient Model

The archetype of positional information mechanisms is the Gradient Model (fig. 2a) [486, 798, 858, 881, 996, 997, 1005]. In the familiar 'source-sink' version [24, 149, 996, 997] the axioms are as follows. A diffusible chemical 'morphogen' (signaling molecule) is produced at one end of an array (the source) and consumed at the other (the sink). When a steady state is reached, the concentration has a gradient profile. Each cell records the concentration at its location as a coordinate and 'interprets' this 'positional value' as a state of differentiation. In figure 2a, the rule is that odd-numbered cells become black and even-numbered cells become

---

Fig. 2. Positional information mechanisms. Positional information models assume that cells know where they are (relative to organ boundaries) and are capable of choosing particular states of differentiation based upon this 'area code' [277]. *a* Source-Sink Gradient Model [997]. Circles denote individual cells, and gray circles containing question marks are naive (uncommitted) cells. A diffusible chemical signal or 'morphogen' (M) is produced by the source (6) and consumed by the sink (1), resulting in a steady state where the morphogen concentration ranges linearly between them (from 1 to 6 units), thus describing a 'gradient' profile (stippled triangle). The intervening cells record the concentrations at their locations as 'positional values' (2–5), which they will retain even after the morphogen disappears. Ultimately, they 'interpret' their values as black (odd numbers) or

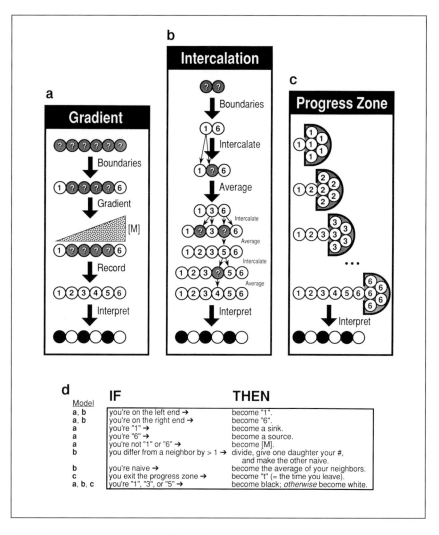

white (even numbers) states of differentiation, though it is unknown whether cells can actually compute 'even' vs. 'odd' (see text). *b* Intercalation mechanism, based upon the Polar Coordinate Model [242, 245]. The outer coordinates are established first. Numerical gaps cause mitosis. Mitoses (thin arrows) are asymmetric since one daughter retains the parental number while the 'intercalated' daughter does not. Naive cells compute a coordinate by averaging the coordinates of immediate neighbors. *c* Progress Zone Model [867, 868]. The gray semicircle represents the 'progress zone' where positional values increase incrementally. Extruded cells keep their values. *d* Conditional 'IF/THEN' rules (phrased in the second person as cellular commands) implicit in one or more of the models depicted above.

white, though, as discussed below, it is uncertain whether cells can actually perform such computations. The utility of the model is that it easily explains how patterns can regenerate after parts have been removed: either (1) cellular memories are erased and the mechanism starts over with fewer cells ('morphallactic regulation'), or (2) mitoses at the cut edge produce cells having successively lower coordinates ('epimorphic regulation') [997].

The precepts of this model were based upon regulative phenomena in numerous developing systems, and its predictions have subsequently been tested in many of those same systems [1008]. Proof of its operation has been adduced for the anterior-posterior axis of *Drosophila* embryos, where the anterior morphogen has been identified as a DNA-binding protein encoded by the *bicoid* gene [187, 243, 649, 854, 856]. The first *periodic* patterns to arise along this axis are 7-striped arrays within which individual 'pair-rule' segmentation genes are transcribed [5, 397, 410]. For each gene, the stripes manifest a 2-segment periodicity (hence the name '*pair-rule*'), but the arrays for different genes are out of phase relative to one another, permitting an overlapping combinatorial code for cell states [269, 411, 492, 745]. Pair-rule gene expression is controlled via an intermediate echelon of 'gap' genes, so-named because mutant larvae are missing multisegmental swaths of cuticle [282]. Moreover, each stripe of every pair-rule gene appears to be under the control of a different set of maternal, gap, and other pair-rule genes [398, 668, 669, 722, 799, 815, 937], implying that this periodic pattern is merely an illusion created by a highly *aperiodic* mechanism [6]. However, other models (to be discussed later) challenge this conclusion.

In other versions of the Gradient Model, various amendments have been proposed: (1) all cells act as weak sinks, instead of one end acting as a strong sink [573, 866]; (2) all cells act as sources and establish a gradient by pumping the morphogen in one direction [134, 278, 486, 997]; (3) there are two opposing source-sink gradients, and cells measure the ratio of the two different morphogens [14, 156, 157, 395, 744, 997, 998]; (4) the concentration gradient is not created by sources or sinks but by a reaction-diffusion mechanism [279] or by a cell-signaling process that is formally equivalent to reaction-diffusion [22]; (5) instead of a concentration gradient, cells differ in the degree to which two oscillating chemical reactions are out of phase [124, 144, 310]; and (6) a gradient stage is avoided altogether by having two opposing wavefronts of cell-surface interactions directly establish a step-function concentration profile for the morphogen [566].

If an organ regulates epimorphically, then removal of the gradient's high point should cause the remaining tissue to produce a mirror-image duplicate of itself, whereas excision of areas lacking the high point should allow complete regeneration [441]. Hence, the high point should be locatable by systematically amputating various parts of an organ. When such experiments were performed with the developing wing of *Drosophila*, all four quadrants underwent duplication [245]. None regenerated. This paradoxical inability to find a high point led to a new model where positional information is not specified by conventional gradients.

The Polar Coordinate Model

Unlike the Gradient Model, the Polar Coordinate Model [81, 245] assumes that cells assess their positions by observing the coordinates of their immediate neighbors (presumably by contact between their cell surfaces [242]), rather than via a long-range diffusible signal. The model invokes polar (radial and angular) instead of Cartesian (perpendicular gradient) coordinates. Excision of a sector supposedly leads to (1) healing together of normally nonadjacent cells; (2) local proliferation in response to the positional disparity; and (3) an 'intercalation' zone which bridges the gap of coordinates via the shorter of the two possible routes (clockwise or counterclockwise) around the circumference. Pieces containing less than half of the circumference (e.g. a quadrant) would therefore duplicate, regardless of their location in the organ as a whole (i.e. there would be no high point). Interactions between developing limb buds and regenerating limb blastemas in amphibians suggest that the same mechanism is used for both the development of the original pattern and its regulative responses to surgical manipulations [612–614].

How would the coordinate system arise during normal development? If the peripheral coordinates are established first, then newborn cells could adopt intermediate coordinates until all gaps have been eliminated (fig. 2b) [153, 242, 245, 614]. An intercalation mechanism can explain several phenomena which the basic Gradient Model cannot, including: (1) why some organs never exceed a definite size even when given additional time to grow (because growth should stop automatically when all coordinates are present; *double*-gradient models can also explain determinate growth) [14, 79, 80, 245, 997]; (2) why defects in cell adhesion can lead to overgrowth (because adhesion should be crucial for contact-mediated communication between adjacent cells, less so for diffusible signals) [78, 427, 548]; and (3) why cell death during the growth phase can cause pattern duplications (because removal of more than half an organ should cause the remaining piece to duplicate at *any* stage, not merely after growth has ceased and all coordinates have been specified) [75, 286, 287, 420, 690, 694, 749, 790].

Tests of the Polar Coordinate Model in insect, salamander, and chick limbs have yielded extensive supportive evidence, implying that vertebrates and arthropods use a common strategy to construct their appendages [81, 415, 421]. However, many regulative properties of chick wing buds can also been interpreted in favor of a gradient mechanism [866], raising doubts about the monophyly of the process, and questions remain about other phenomena which the model cannot easily explain [242, 441].

In an effort to alleviate some of the model's shortcomings, other authors have proposed (1) specifying the angular and radial coordinates via the ratio of two morphogens [1, 786]; (2) retaining the rules for intercalation but computing positional disparities from Cartesian coordinates [153, 305, 441, 445, 448, 747, 977]; (3) partitioning the coordinate system into discrete territories which intercalate only when an entire territory is missing [311, 575, 785, 793]; or (4) dispensing with coordinates altogether and using a 'smoothing' rule to control intercalation [509, 510, 597, 977, 978].

The Progress Zone Model

Though sometimes categorized as a gradient model [1008, 1009], the Progress Zone Model is uniquely different from any of the mechanisms discussed thus far. It emerged from experiments on the wing rudiments of chick embryos. The chick wing develops from a bud which grows mainly in a 'progress zone' at its tip. Reciprocal grafts between young and old buds led to the idea that the proximo-distal coordinate of a cell reflects the length of time that it spends in this zone [867, 868]. The cells would thus acquire positional information via temporal information (they presumably can measure time and stop their 'clocks' when they exit the zone; fig. 2c) in contrast to conventional gradient models (where positions are specified independently of growth) and intercalation mechanisms (where signaling is correlated with growth but not with time per se).

Puzzles and Paradoxes

Can Cells Actually Perform Mathematical Calculations?
All positional-information models assume that cells record their positions as a numerical quantity ('positional value') [997, 1004]. Epimorphic regulation requires that new numerical states be computed from old ones, and the Shortest Intercalation Rule [245] demands that cells choose the smaller of two numbers. Whether cells can actually perform such computations is an interesting question. Precedents do exist for elementary mathematical capabilities in some cells. Thus, animal neurons can compute the sum of positive and negative inputs via effects on their membrane potential [525, 724], and *Drosophila* and nematode cells can assess the ratio of X chromosomes to autosomes during sex determination [382]. However, some patterning models require cells to compute cosines or even more complicated functions [153, 747, 786]. Must cells have the equivalent of a secondary school education in order to participate in pattern formation? Chapter 7 explores the limited abilities of embryonic cells to store and process information, and the general conclusion is that cellular 'intelligence' is closer to a kindergarten level.

How Does Interpretation Work?
The issue of cell 'intelligence' is especially troubling with regard to how cells interpret positional information [134, 1008]. For the Gradient Model, the orthodox view is that morphogen-sensing genes have different concentration thresholds that are sharpened by autocatalysis and intergenic repression [14, 34, 292, 456, 512, 540, 572, 577]. Evidence supporting this view has come from studies of *Drosophila* gap genes [402, 474, 499, 854] and *Xenopus* mesoderm inducers [322, 323]. (Unorthodox interpretive schemes have been proposed by Babloyantz [23, 24] and Goodwin [306].) However, the level of detail in most anatomical patterns is orders of magnitude greater than a French Flag (or a striped *Drosophila* embryo) and the theoretical precision of concentration sensing is crude by comparison [136]. Conceivably, organs could be subdivided into hierarchies of nested gradients [134, 321, 558, 997], permitting a serial combinatorial code of positional values, but decoding would still be a problem [1008, 1011]. To appreciate the dilemma, consider that each of the million-or-so hairs in your skin would have to possess a unique 'area code' [277] and decipher its code by looking it up in the genetic equivalent of an enormous 'area-code/ differentiated-state' directory [134, 138, 146, 540, 706] (cf. Beardsley [48]

and Davidson [162]). A similar dilemma pesters the cognitive maps that are supposed to control human limb movements [21, 61].

Collective Amnesia?
An embryo can theoretically use a single coordinate system to specify anatomical patterns in many different organs by merely changing the rules for how the coordinates are interpreted [997]. In that case, cells would need to identify their 'organ state' (e.g. leg vs. wing) before deciding how to translate their coordinates into cellular states of differentiation (e.g. neuron vs. myocyte). Mutations in genes that encode organ states should cause the cells of one organ to construct an anatomy typical of another, as opposed to *cell type* interconversions [63, 65, 92, 369, 599, 652, 891, 916, 938]. Indeed, many such 'homeotic' mutations have been found in both plants [67, 97, 120, 221, 469, 587, 773] and animals [49, 50, 66, 118, 784], including man [46, 419, 797]. They have been studied intensively in *Drosophila* [7, 450, 488, 547, 666]. For example, mutations in several *Drosophila* genes can cause a partial transformation of the antenna into a second leg. Strangely, a clonal analysis of leg-tissue islands in the antennae of *Antennapedia* flies showed that the transformations occur concomitantly in groups of neighboring cells, i.e. via *proximity,* not via *pedigree* [692]. Similarly, another type of homeotic transformation (termed 'transdetermination' [339]), which is routinely encountered during long-term culture of nonmutant *Drosophila* tissues, also affects nonclonal groups of cells [263] (cf. Karlsson [432] for distinctions between transdetermination and ordinary homeosis). The perplexing implication is that organ states such as 'legness' or 'wingness' may be not be properties of single cells, but rather of cell *clusters* (cf. Chandebois [105]). *Xenopus* muscle differentiation likewise appears to involve a 'community effect': a cell will only shift its fate (in response to an inducing signal) if a sufficient number of its neighbors also does so [334].

The Antenna-Leg Paradox
Within the *Antennapedia* antenna, specific leg structures always develop in predictable positions, allowing a mapping of corresponding antennal and leg domains [265, 693]. The map has been construed as evidence that the antenna and second leg interpret the *same* coordinate system according to different sets of rules (cf. Haynie and Bryant [362]). Given this apparent homology, however, their regulative behavior is difficult to understand. In *Drosophila* the external structures of the head and thorax develop from separate pockets of epidermis called 'imaginal discs' [266]. The second leg

develops from its own disc, but the antenna comes from part of a disc that also forms the eye. When the eye-antennal disc is bisected, the eye portion regenerates an antenna, and the antennal portion duplicates [262, 264], implying – according to the Polar Coordinate Model – that the antenna comprises *less than half* of a larger eye-antennal regulative 'field' [408, 929]. By contrast, when parts of the second-leg disc are removed, it behaves as an *entire* field [283, 749]. How can less than half of a coordinate system equal a whole coordinate system? Conceivably, the eye may constitute an outer annulus of the coordinate system, rather than an oversize sector. A related riddle is: if the arista (the feathery tip of the antenna) is homologous to the tarsus according to the *Antp* map and a comparable map inferable from *spineless-aristapedia* phenotypes [86, 851], then why does the boundary line separating the anterior and posterior 'compartments' [151] of the eye-antennal disc bypass the arista entirely [604] (cf. Brower [73]), whereas it bisects the tarsus all the way down to the claws [493, 824]? The importance of this question stems from the supposed significance of the anterior-posterior compartment boundary in general [151, 491, 556] and its presumptive role in *causing* appendage outgrowths in particular [126, 267, 574, 576, 578].

Warped Coordinate Systems?

Another peculiarity of *Drosophila* leg discs is that, unlike the wing discs, one quadrant (the upper medial one) can regenerate the remaining three-quarters of the disc [769]. The Polar Coordinate Model explains this oddity by assuming that more than half of the leg's circumferential coordinates are crowded into this quadrant [245]. The difficulty with this ad hoc remedy, however, is that it creates a new problem. If growth stops when all discrepancies between adjacent positional values have been eliminated [80], then how can one part of an organ acquire a three-fold higher density of coordinates?

How Do Body and Limb Coordinate Systems Mesh?

The *Drosophila* embryo apparently uses a Cartesian (anterior-posterior and dorso-ventral) coordinate system [649], but the imaginal discs, which arise as inpocketings of the embryonic body wall [44, 123] use a polar coordinate system [245]. Do disc cells algebraically convert one system into the other, or do they erase their positional memories and create a polar system de novo [242]? (The same dilemma applies to body vs. limb axes in amphibia [441].) Attempts to reconcile the two coordinate-system models [267, 575, 578, 786, 971] have not been wholly satisfying.

Chapter 2:
Prepattern Mechanisms

Models Involving Physical Forces

Physical forces can create periodic patterns in inert matter, e.g. the rings of Saturn (gravity); ocean waves, sand dunes, mackerel clouds (fluid mechanics); and the harmonic waveforms of musical instruments (vibrations) [16, 230, 703, 734, 735, 949]. The physical properties of cells and their extracellular materials (e.g. viscosity and elasticity) can also produce local deformations in response to internal or applied forces [30, 36, 288, 351, 619, 637, 650, 661, 662, 944, 950], and these distortions could theoretically promote the development of structures.

Before the ascendancy of the positional information paradigm, the dominant idea in the field of pattern formation was the concept of 'prepatterns' proposed by Curt Stern [831, 832] in 1954. Stern imagined that epithelial folds could cause 'stress points' (fig. 3a) where structures such as bristles might be induced:

> The larval imaginal discs (of *Drosophila*) are made up of cell layers that are folded in complex ways. Let us postulate that differentiation of bristles occurs at those points at which folds cut across each other. According to this hypothesis, an allele that leads to differentiation of a specific bristle would be involved in provoking the formation of specific folds. Another allele whose phenotypic effect does not include formation of the bristle would be responsible for a different kind of folding of the imaginal disc. The different types of folding of the discs would constitute different patterns. Since these patterns would precede the appearance of their corresponding bristle patterns, I refer to them as prepatterns [835].

The foundations for the prepattern hypothesis were laid by Stern's mentor Richard Goldschmidt, who (1) showed that some color patterns on lepidopteran wings are preceded by latent prepatterns in the rates of wing

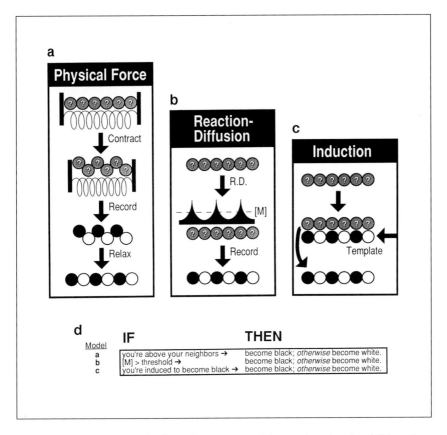

Fig. 3. Prepattern mechanisms. Prepattern models postulate that the visible pattern of structural elements (in this case, black cells) is preceded by a nearly congruent ('isomorphic') array ('prepattern') of sites ('singularities') which prompt the cells to differentiate as pattern elements. Singularities may differ from the background either physically or chemically (or both). *a* Physical Force Model, based upon Stern's original notion of prepatterns [835]. A layer of cells is compressed by an external force (schematically depicted as a spring) which causes it to buckle. The stresses at the apices of the folds then function as singularities, causing the cells there to become black. (Buckling forces have been implicated in causing corrugations in the cerebral cortex [495], in the ciliary body of the bird eye [37, 38], and in insect tracheae [959] and tarsi [931].) *b* Reaction-Diffusion Model, of the kind proposed by Turing [907]. Chemically reactive molecules diffusing at different rates cause the concentration of a product – the 'morphogen' (M) – to peak at 'wavelength' intervals. Wherever the concentration exceeds a threshold (dashed line), the cell at that site becomes black (cf. Waddington [931]). *c* Induction Model, analogous to neural induction in vertebrates. An underlying 'template' pattern induces corresponding states of differentiation in an apposed layer of naive cells. *d* Conditional 'IF/THEN' rules implicit in the models depicted above.

scale maturation [293, 294, 803]; (2) reviewed the literature on pattern formation in 1938, including a conjecture by Krieg (1922) that the striped arrangements of hair follicles in the tiger embryo (another example of a prepattern) are attributable to 'tensions within the skin at the time of pattern determination [294]; and (3) speculated on a primary role for 'growth tensions' in tissue patterning [294].

Because a prepattern supposedly arises from forces within a whole tissue, it should be resistant to local perturbations. Thus, if a mutation alters a prepattern, then a few mutant cells in a wild-type background should not disturb the overall (wild-type) pattern but rather should acquiesce 'nonautonomously' [609, 859] in its formation. Given this logic, Stern was surprised when the first *Drosophila* mutation that he tested in this way behaved autonomously. In homozygous flies, the mutation *achaete* causes the absence of a specific thoracic bristle. In genetically mosaic flies where most of the thorax consists of wild-type tissue, *achaete* cells typically fail to form a bristle when they reside at that site – in effect ignoring the surrounding majority of cells [831]. From this result Stern reasoned that prepatterns cannot be sufficient for the induction of structures: cells must also be 'competent' to respond to signals from the prepattern, and *achaete* must be affecting the competence of cells at a specific site. For example, cells might have a threshold of strain above which they become bristles, and *achaete* raises the threshold at one location. When more than a dozen other pattern-affecting mutations were similarly tested in genetic mosaics, most were found to also behave autonomously [888]. Among them was a homeotic mutation – *spineless-aristapedia (ss^a)* – which (like *Antennapedia*) causes a partial transformation of antenna into leg. According to the prepattern hypothesis, the autonomy of *ss^a* indicates a hidden prepattern for leg structures in the developing antenna (to which only the mutant cells can respond) [691, 727], and because other homeotic mutations also manifest autonomy [505, 506, 603, 887, 889], one is led to the absurd inference that each imaginal disc must contain a hidden prepattern for every other disc [441]. Positional information provides a more plausible hypothesis since such mutations could simply be altering the rules by which an invariant system of coordinates is interpreted [888, 997]. Consequently, prepattern models have waned as explanations for adult cuticular patterns. However, they have recently experienced a revival vis-à-vis embryonic body segmentation because of reaction-diffusion schemes that can explain the striped patterns of segmentation gene expression [353, 478, 480, 539, 624, 625].

Reaction-Diffusion Models

Two years before Stern published his hypothesis, Alan Turing, a founder of modern computing [381], described a clever model based upon *chemical* prepatterns [907]. By postulating imaginary chemical reactions between molecules which diffuse at different rates, Turing showed that a homogeneous distribution of the molecules is unstable under certain conditions [219, 615]. Statistical fluctuations become amplified into peaks and troughs of concentration (fig. 3b) at wavelengths which depend upon the relative rates of reaction and diffusion. Only recently have actual chemical reactions been found which do indeed behave in this way [500, 667, 689, 981]. In two dimensions, Turing's equations can produce periodic patterns reminiscent of zebra stripes or leopard spots [31]. Virtually any periodic pattern in any dimensional space can be simulated by the reaction-diffusion schemes of Alfred Gierer and Hans Meinhardt [279, 570, 571, 573, 579, 580] and others [71, 619, 625, 626, 642, 643], who have modified the parameters of catalysis and diffusion in Turing's model and added new assumptions. Generically, these types of models predict certain pattern modulations as a function of shape [288], and the predictions are strikingly confirmed in the coat markings of mammals [32, 33, 616–618] and the fruiting buds of slime molds [90, 567, 568].

A curious property of reaction-diffusion models is 'stochastic indeterminacy' [313, 642, 643, 1013]: the final configuration of the pattern elements cannot be exactly predicted from the starting conditions. (Positional information models are deterministic: they should yield identical patterns from trial to trial.) Indeed, many anatomical patterns that manifest a certain regularity at one level are indeterminate at another level [948, 955]. For instance, human fingerprint patterns manifest a uniform ridge spacing, but ridge configurations are not identical in identical twins [154] (cf. freckles). Such patterns are 'epigenetic' [408] insofar as their features are not specified genetically. Presumably, the genes merely establish the starting conditions (e.g. reactant concentrations), with the outcome being dictated by the same random perturbations (e.g. concentration fluctuations) that initiate the process [307, 308]. Mechanochemical models, which combine aspects of both physical-force and reaction-diffusion mechanisms, have been designed and, in some cases, augmented with further assumptions to make them less indeterminate [288, 619, 660, 663].

Induction Across Layers

A trivial explanation for the origin of a pattern is that it is imprinted from a 'template' prepattern in an apposed layer of cells (fig. 3c) [105, 106, 409, 433, 434, 754, 927, 928, 1001]. Such inductions are common in vertebrate skin. Thus, bird feathers and mammalian hair are primarily epidermal in construction, but their positions are determined by underlying clusters of dermal cells [776, 805]. In *Drosophila* wings, the veins in the ventral layer are induced by those in the dorsal layer [251]. A peculiar variation on this theme is the long-distance induction of 'neurobarrels' in the rat trigeminal system by afferents from the facial whiskers [767]: the whisker-vs.-neurobarrel patterns are remarkably isomorphic, and the cautery of particular rows of whisker follicles in a neonate causes a rostrocaudal cascade of abnormalities in the corresponding neurobarrels of the brainstem, thalamus, and somatosensory cortex [45, 457]. In an analogous manner, optic cartridges of second-order laminar neurons are induced by afferent retinal axons in *Drosophila* [586] and other insects [569].

The Preformationist Paradox
The induction of one pattern by another begs the question: How does the inducing pattern arise? Indeed, the entire prepattern school of thought has been criticized for implying an infinite regression of patterns induced by prepatterns [756, 803, 927, 997]. Historically, this criticism was justifiably leveled against the antiquated notion of preformationism, which argued that eggs (or sperm) contain preformed homunculi which, in turn, must have eggs bearing smaller homunculi ad infinitum [408, 602, 655]. The objection would be legitimate in this case if prepatterns could only be established by induction, but they can also arise de novo via physical forces or chemical reactions as discussed above. Comparable misunderstandings have repeatedly arisen from a failure to appreciate the cardinal distinction between the rules that generate a pattern and the information content of the pattern itself [14, 15, 69, 504, 826, 828, 1010].

Prepatterns vs. Positional Information

All prepattern models employ 'singularities' [834, 835] (sites where physical or chemical parameters differ from the background) as cues for inducing structures. A structure should form wherever a singularity is present unless certain cells are unable to respond. Thus, a prepattern and its subsequent pattern can be isomorphic [625], or the prepattern may have extra 'cryptic' singularities [561, 833]. Prepattern models differ from positional information models in several key respects [136, 625, 888, 998, 999, 1007, 1008]:

(1) *Identical structures use identical signals.* For a pattern of 3 bristles, there would be 3 identical singularities, all of which would directly signify 'Make a bristle', so the cells would never know their positions. In contrast, positional information would use 3 different positional signals as bristle commands – e.g. 'You're at (x_1, y_1)', 'You're at (x_2, y_2)', and 'You're at (x_3, y_3)' – and hence the cells would know where they are.

(2) *Cells can be 'stupid'.* If a cell is located at a prepattern singularity, then all it must do is switch its state relative to the cells of the background. There is no true interpretation stage: a nudge suffices. With positional information, however, cells must not only be 'bilingual' (able to translate positional coordinates into differentiated states) but their vocabulary must be as large as the number of pattern elements, since each location uses a different signal even if the elements are identical. The distinction is analogous to bitmaps vs. vector representations in computer graphics: in bitmaps (= coordinate systems) the states of all pixels (= cells) must be specified regardless of the type of image, whereas most geometric patterns can more economically be encoded by a vector format (= prepattern) [673, 726].

(3) *The number of pattern elements is size-dependent.* Whereas coordinate-system models are designed to ensure pattern constancy regardless of pattern size, prepattern mechanisms inherently lack this ability [558, 621], though ad hoc amendments can be added to enable them to do so [24, 352, 665, 670, 671]. Since size can usually be altered easily, the demonstration of size-dependence in a given system can serve as a convenient operational criterion for ruling out the sole involvement of positional information mechanisms. It does not prove a prepattern mechanism, however, since other types of mechanisms, e.g. Darwinian ones, are also size-dependent. Because absolute size is a function of both cell size and cell number, it is possible to vary cell size (e.g. by polyploidy) and cell number (e.g. by starvation) separately to observe whether the pattern responds to either or both of these factors. Structures which vary in number as a function of organ size, cell size, or cell number include: zebra stripes [32, 33], *Drosophila* bristles [367, 758], *Hydra* tentacles [64], melanophores and ciliated epidermal cells in *Bombina* (a frog) [211, 212], whorls of fruiting bodies in

Polysphondylium (a cellular slime mold) [809], wing veins in *Ephestia* (a moth) [559], pigment stripes in alligators [620], and ocular dominance columns [894] and lateral line primordia [984] in *Xenopus*. Notable patterns that do *not* change with cell size are the number of somites in *Xenopus* [133, 135, 346] and the number of '*ftz*' stripes in *Drosophila* embryos [863].

Hybrid Models

The greatest strength of positional information mechanisms is their regulative ability; their greatest weakness is the amount of information processing they require [134, 1008]. Prepattern mechanisms have a complementary strength and weakness. It was inevitable, therefore, that hybrid mechanisms would be proposed which exploit the best features of the two types of models [1007]. Examples include:

The Gradient/Reaction-Diffusion 'Superposition' Model [573]
In many periodic patterns the structures are similar but not identical. For example, your fingers resemble one another but differ in length. Finger positions could be designated by a reaction-diffusion prepattern, with each finger growing to a different length based upon a gradient along the distal edge of the palm [134, 138, 1008]. Hybrid models of this kind are economical because (1) positional information need only be interpreted by a few rudiments (thereby minimizing information processing), and (2) positional signals need only be accurate enough to distinguish the rudiments (thereby minimizing the demands on the signal-to-noise ratio). The slight differences among human fingers would not require any interpretation of positional information per se since they could arise directly from a gradient in tissue growth (cf. Child [110], Huxley [407], and Thompson [882]), but there are many instances where one member of an anatomical series is greatly exaggerated, e.g. the wing strut of the pterodactyl (an enormous fourth digit), elephant tusks (enlarged incisors), and the 'sabers' (canines) of saber-toothed tigers. Such cases have been marshaled as evidence for a 'Principle of Non-equivalence' [508, 515, 1000], which postulates that (1) all cells have unique positional values, and (2) the ability to change the interpretation of those values genetically allows the independent evolution of formerly identical structures. Hybrid models, while not violating this principle, demand that any changes in the number or arrangement of struc-

tures must be explained otherwise – namely, in terms of the prepattern portion of the mechanism.

The Gradient/Reaction-Diffusion 'Tuning' Model

The network of genetic interactions that governs the expression of *Drosophila*'s pair-rule segmentation genes appears to be highly complex – involving both positive and negative signals, reciprocal and nonreciprocal interactions, redundant functions, and stripe-specific sets of enhancer elements [5, 47, 98, 99, 350, 397, 410, 413, 675, 774]. However, Lacalli and Harrison [479] have argued that some of the complexity may be illusory: pair-rule gene products may be participating in reaction-diffusion mechanisms whose parameters (e.g. diffusion rates) are merely 'tuned' by gap gene products. The notion that diffusion rates are important for striping is supported by the polarized release of pair-rule gene transcripts from the apical ends of syncytial blastoderm nuclei, where the translated proteins would experience markedly greater diffusional impedances than gap gene proteins, which are not so confined [165].

The Progress-Zone/Oscillator Model

A peculiar feature of the development of the chick wing is that the time required for the specification of each bone rudiment within the progress zone is uniform despite a huge range in eventual lengths (e.g. wrist elements vs. humerus) [513, 514, 542, 867, 868]. The surprising implication is that the limb skeleton may begin as a periodic pattern of identical elements that diverge in size through subsequent growth. Wolpert and Stein [1011] have devised a hybrid model (cf. Meinhardt [574]) where oscillating chemical reactions within the progress zone produce a periodic prepattern of concentration peaks as cells leave the zone. If each peak becomes a bone, then the different shapes and sizes of the bones could be controlled by the duration of time spent in the zone, as in the original Progress Zone Model. Thus, the rudiments would be created by a prepattern, with positional information steering them into different fates, as in the Superposition Model above. Analogous cases where a meristematic growth zone undergoes cyclic changes – leaving a periodic pattern in its wake [949] – include: tree rings [347] (alternating xylem and phloem), *Xenopus* tailbud somites [133], terminal segments in short-germ insects [244, 755, 757, 770], mouse molar teeth [538], barred feathers [294, 601, 641, 917], agouti hairs [294], and the spiral series of chambers in nautilus shells [132, 882, 935].

Chapter 3:
Determination Wave Mechanisms

Antedating the prepattern (1954–1969) and positional information (1969–present) epochs in developmental biology was a period (1920–1954) when the idea of 'determination waves' (a.k.a. 'determination streams' or 'spreading fields') was paramount [294, 338, 803]. The concept was a scion of Boveri's 'gradients' and Spemann's 'organizer' [134, 803]. It was conceived by Richard Goldschmidt and expounded in 1920 [293, 294]. He envisioned a propagating signal or substance that spreads from an 'organizing center' to control the fate of every cell that it reaches. The hypothesis was buttressed by demonstrable waves of mitosis or pigmentation in various organisms [803], and it gained notoriety through the experiments of Kühn and his collaborators on wing coloration in the flour moth *Ephestia kühniella* [294, 338, 475, 646]. Each forewing of this moth has two parallel bands of white scales, and the positions of the bands can be shifted through microcautery or heat shock. The earlier the interference, the more the bands recede toward two points on the wing margin – as if they are wavefronts that spread from those centers [294, 476] (but cf. Toussaint and French [896] for contrary evidence). Periodic ('rhythmic') bands in lepidopteran wings have likewise been explained in terms of oscillating chemical wavefronts [294, 803, 949].

Like prepattern models, wave models employ identical signals for identical structures, but because pattern elements are established sequentially, the earlier ones need not 'wait' for the later ones before they differentiate. Thus an isomorphic prepattern sensu stricto need never exist. Because morphogenesis and differentiation proceed serially along definite axes in so many diverse organs and organisms [13, 136, 418, 524, 658, 659, 714, 775, 837, 957, 1020], it seems likely that determination waves control pattern formation in at least some of these systems.

Chemical Waves

Propagated chemical reactions can form patterns of parallel, concentric, or spiral stripes which resemble the outcomes of Turing mechanisms [733]. Among them are the Belousov-Zhabotinsky reaction, where the concentration of a chemical oscillates in time and space [227, 470, 610, 619, 643, 976, 979, 980], and the Liesegang reaction, where the diffusion of one electrolyte into a gel containing another electrolyte causes a salt product to precipitate in alternating bands [364, 373, 611, 919]. Periodic waves of Ca^{2+} release have been documented in monolayer cultures of glial cells [107, 147], inside oocytes [497, 806], and in other types of cells [584], but whether ionic periodicities of this sort play a causative role in tissue patterning remains speculative [294, 364, 389, 803, 927, 949]. Propagated waves of cyclic AMP control the aggregation of slime mold amoebae but apparently do not assign the cells a state of differentiation (i.e. prestalk vs. prespore) [297].

Cellular Automata

Interestingly, the Belousov-Zhabotinsky reaction and other reaction-diffusion mechanisms [22] can be simulated by employing contact-mediated cellular communication, instead of diffusible chemicals [173, 270, 544, 545, 551, 982]. Generally speaking, such 'cellular automata' models postulate arrays within which each cell can exist in a finite number of states, and the state of a cell at time '$t + 1$' is determined by the states of its neighbors at time 't' according to predefined rules that apply to the entire array [119, 255, 337, 360, 706, 886, 924, 988, 989]. Given particular rules and starting conditions, surprisingly intricate patterns can emerge and propagate across the array [358]. This genre of models was popularized in the early 1970s by John Conway's clever game 'Life', where cells live or die or are reborn, depending upon how many of their neighbors are alive [26, 254–256]. Three types of automata schemes, which utilize different kinds of determination waves, are discussed below.

The Sequential Induction Model

The gross anatomy of the vertebrate body is constructed by branched chains of inductive events, beginning with the notochordal induction of the central nervous system and followed – in the case of the eye for example – by the optic cup inducing the lens and then the lens inducing the cornea [341, 644, 807]. Comparable cascades may establish the states of *individual* cells within tissue monolayers. The clearest demonstration is the development of the *Drosophila* retina, where the inductive cascade within each cluster of 8 photoreceptors begins with photoreceptor R8 and ends with R7 [340]. Two different 'recruitment' scenarios are consistent with the available data [42, 986]: (1) The inductions might only involve one inducing cell at a time, as in the scheme ('→' denotes an induction) 'R8→R2, R2→[R3 and, later, R1], R8→R5, R5→[R4 and, later, R6], and R8→R7' [28]; or (2) a *combination* of signals from several adjacent cells may be required, according to rules such as 'IF you are next to R1 *and* R6 *and* R8, THEN become an R7' [741, 890, 892, 893], and this 'epigenetic combinatorial code' [713] could extend to other ommatidial cell types (pigment cells, cone cells, etc.) as well [91, 92, 715]. In either case, the inductions do not iterate: the R7 of one cluster does not induce another R8 as the 'seed crystal' for an adjacent cluster [496, 1012], even though the clusters do develop sequentially across the retina. In theory, iterative inductions can construct periodic patterns [105, 106, 580, 581]. For the array depicted in figure 4a, a unidirectionally transmitted signal instructs black cells to induce white neighbors, and vice versa. In two dimensions, either parallel or concentric stripes could arise, depending upon whether the original source of the signal is a line or a point.

In monolayer cultures of *Dictyostelium* amoebae, scattered 'pacemaker' cells emit regular pulses of the diffusible signaling molecule cyclic AMP, which induces nonpacemaker cells to move toward the signal source [171, 172, 600, 635, 707, 729]. The signal is relayed throughout the cell population because each cell emits its own pulse upon stimulation, and reverse-propagation of the wave (i.e. cells responding to reflected signals from outlying cells) is prevented by a transient refractory period. Curiously, stripes (either concentric or spiral) of alternately moving vs. stationary cells arise because the refractory period includes two phases: a burst of movement followed by a stationary period [231, 914, 915]. Thus, this alternating pattern is not caused by alternating signals (i.e. moving cells telling neighboring cells not to move and vice versa) but by one signal which leads to an automatic succession of two states within every cell. Such dual wavefronts have been termed 'primary' and 'secondary' because one causes the other but not vice versa [765, 1020].

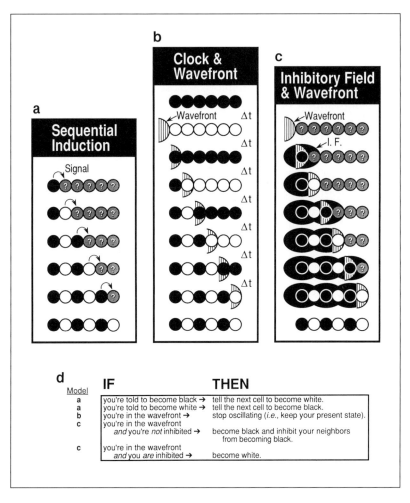

Model	IF	THEN
a	you're told to become black →	tell the next cell to become white.
a	you're told to become white →	tell the next cell to become black.
b	you're in the wavefront →	stop oscillating (*i.e.*, keep your present state).
c	you're in the wavefront *and* you're *not* inhibited →	become black and inhibit your neighbors from becoming black.
c	you're in the wavefront *and* you *are* inhibited →	become white.

Fig. 4. Determination wave mechanisms. Determination wave models create patterns sequentially from one end of a cellular array to the other, utilizing a single 'determination wave' whose propagation mode (and cellular effect) depends upon the assumptions of the particular model. *a* Sequential Induction Model. Alternating signals ('become white' or 'become black') are relayed along the array, causing cells to adopt states unlike their neighbors. *b* Clock and Wavefront Model [146]. All cells synchronously oscillate between black and white states (period = Δt) until a wavefront (striped semicircle) reaches them, causing them to keep the state that they happen to have at the time. *c* Inhibitory Field and Competence Wave Model [196]. A cell is only competent to become black when it resides in the wavefront (striped semicircle). Upon becoming black a cell immediately establishes an inhibitory field ('I.F.', black oval), within which no other cell can become black. *d* Conditional 'IF/THEN' rules implicit in the models depicted above.

The Clock-and-Wavefront Model

Many types of cellular oscillators are known [58, 124, 125, 291, 304, 305, 619, 664, 977], the most familiar being the mitotic cycle. Theoretical mechanisms that explicitly employ such oscillators in patterning include the Phase-Shift Model of Goodwin and Cohen [124, 144, 310], the Clock-and-Wavefront Model of Cooke and Zeeman [134, 146, 1020], and the previously discussed Progress-Zone/Oscillator Model of Wolpert and Stein [1011]. In all of these models, there are two essential components: (1) a two-state 'clock' which oscillates synchronously (or nearly so) in all cells, and (2) a unidirectionally traveling 'wavefront' which is capable of stopping (or being modulated by) the cellular clock. In the Progress-Zone/Oscillator Model, the wavefront is associated with a terminal growth zone, whereas the other two models can operate within static (nongrowing) arrays. If the cells within an array are synchronously oscillating between two states, then a wavefront that stops the oscillations will leave behind a periodic pattern of alternating states. In figure 4b, the wavefront travels at a rate of two cells per oscillation cycle. A faster rate of travel would yield more than one black or white cell in each 'bandwidth' of the final pattern, and the widths of the black vs. the white bands could be made unequal by supposing that the cells spend unequal amounts of time in the two states during each cycle.

Although the wavefront could be propagated by a relayed signal as in the sequential induction mechanism, Cooke and Zeeman [146] present an intriguing alternative: in addition to its oscillator clock, each cell might have what is tantamount to an alarm clock [cf. refs. 12 and 801]. If the array is spanned by a morphogen gradient, and each cell schedules its alarm clock for a time corresponding to its morphogen concentration, then the alarm clocks could later ring in sequence from one end of the array to the other (causing the oscillator clocks to stop and resulting in a periodic pattern) *without any intercellular communication.* Whereas a 'relay wave' can theoretically be blocked surgically (e.g. by removing a cell or inserting a barrier, as one might stop a chain of falling dominoes), a 'schedule wave' cannot [135, 170, 370, 496, 524, 958, 1012]. *Xenopus* somites, which arise sequentially in the typical vertebrate manner, behave according to the schedule wave scenario [143, 215, 679]: (1) the wave cannot be physically halted (no matter how early the operation is performed) [170], and (2) somite number is independent of both body size and cell number (as expected for a gradient mechanism) [133, 135, 224, 346]. In chick embryos, heat shocks cause spatially periodic somite abnormalities that are correlated with the duration of the cell cycle [696, 697], suggesting that somitic oscillators may be mitotically coupled [836]. Other systems where the cell cycle may gate cells into different pathways (cf. Reinert and Holtzer [719]) are *Dictyostelium* (where amoebae become prestalk vs. prespore based upon their cell-cycle phase at the commencement of aggregation [297]), the chick limb bud (where the proximodistal skeletal elements arise at a rate of one element per cell cycle [513]), and the ferret brain (where a cell's cell-cycle phase causes it to migrate to a particular layer of the neocortex [564]). Slack [795] has outlined a hybrid 'Clock-and-Wavefront/Gradient' Model, in which cells progress through an entire 'clockface' of scalar states per cycle (instead of oscillating between only two states), and the effect of the wavefront is to stop each cell's clock, resulting in a sawtooth series of gradients.

Inhibitory Field Models

Since antiquity, gardeners have known that a plant's apical meristem prevents nearby axillary buds from forming other apices. Surgical experiments by Child [110] and others [408, 731] circa 1910–1930 showed that many *animal* organs (e.g. a hydra head or a newt limb bud) can similarly prevent organs of the same type from arising in their immediate vicinity. Moreover, Child used this concept of an 'inhibitory field' to explain a periodic pattern: he argued that colonial hydroids acquire a regular spacing of hydranths along the stolon because a new bud sprouts whenever the tip grows beyond the inhibitory field of the previous hydranth [109] (cf. Plickert [685, 686]). Sir Vincent Wigglesworth was the first (in 1940) to apply the inhibitory field idea to the patterning of structures *within* an organism [963, 964]. He studied the positions of abdominal bristles in a hemipteran insect. Like the hairs on a human forearm, these bristles are spaced fairly regularly but are otherwise randomly arranged. Wigglesworth found that the new bristles at each successive molt arise in the largest gaps within the previous pattern. He conjectured that bristle cells consume a diffusible substance needed for bristle development, so that the first cells able to form new bristles would be those that are farthest from pre-existing sites.

In contrast to Wigglesworth's imaginary factor, which would act as a *positive* regulator, most subsequent 'lateral inhibition' [280, 579, 660, 788, 839] models postulate inhibitor molecules that would function as *negative* regulators. Although the production of an inhibitor is formally equivalent to

Table 2. Evidence for inhibitory field mechanisms in the patterning of *Anabaena* heterocysts and *Drosophila* bristles

Feature	*Anabaena* heterocysts	*Drosophila* bristles
Specialized ('*S*') cell	Heterocyst (nitrogen-fixing, non-mitotic) [357, 847, 913, 992]	Bristle mother cell, which divides to produce a 4-cell (mechanosensory) or 8-cell (chemosensory) bristle organ [355, 451]
Background ('*B*') cells	Vegetative cells (photosynthesizing, mitotic) [814, 991]	Epidermal cells (nonsensory) [688]
Dimensions of the pattern	One-dimensional (filament) [814]	Two-dimensional (monolayer) [688]
Arrangement of *S* cells	Evenly spaced [995]	Some bristles are evenly ('isotropically') spaced; others are arranged in rows or aperiodic 'constellation' patterns (fig. 5) [369]
Number of *B* cells in each *S* cell interval	Approximately 10 [596, 967]	Approximately 5–10 [355, 367]
Frequency of 'incipient doublets' (pairs of *S* cells that commence development closer than a normal interval)	4% [967]	Occasional [400]
Outcome of 'sibling rivalry' between members of the incipient doublet	One of the 'proheterocysts' completes differentiation; the other regresses to a vegetative state [596, 967, 968]	One of the 'probristle' cells completes differentiation; the other presumably regresses to an epidermal state [400]
Method used to artificially suppress *S* cell development	Puncture of proheterocyst [596] or breakage of filament [967, 990]	Construction of genetic mosaics whose mutant tissue cannot form bristles [835]
Effect of suppressing *S* cell development	Nearby vegetative cell commences heterocyst differentiation [596]	Nearby epidermal cell commences bristle differentiation [831–834]
Genes whose mutant alleles (or excess dosage) reduce *S* intervals	*hetR* [83], *Multiple 7* [969], *Terminal 7/4* [969]	*Delta, Enhancer-of-split, Notch, polychaetoid, scabrous, shaggy, shibire* [103, 369, 788]
Chemicals or other conditions which increase or decrease *S* intervals	Increase spacing: fixed-nitrogen compounds [913, 995] Decrease spacing: 7-azatryptophan [595], rifampicin [995]	Increase spacing: triploidy [367] Decrease spacing: haploidy [758]
Putative inhibitor	Glutamine or glutamine derivative [83, 993] (but cf. Wolk [994])	Product of the *Delta* [365] (or *scabrous* [27]) gene
Putative mode of inhibition	Diffusion [83]	Cell contact [788] (or diffusion [27, 598]
Additional mechanisms supposedly involved in the patterning of *S* cells	Cell-lineage rule: each vegetative mitosis is asymmetric, and only the smaller daughter can become a heterocyst [594, 967]	Equivalence groups: only cells in certain areas can become bristles [369, 789]

the consumption of an inducer in terms of the types of patterns that are formed [114], a failure of the mechanism (e.g. caused by mutations) would lead to different 'default' outcomes: either complete absence or ubiquitous differentiation of the pattern elements, respectively. Both lateral inhibition and 'medium-range' (distances of several cell diameters) activation [580] are thought to control vulval cell fate in *Caenorhabditis elegans* [327] and the expression of 'segment-polarity' genes in *Drosophila* [376, 554]. Inhibitory fields need not be mediated by diffusible chemicals: they could theoretically be caused by mechanical forces [324, 325], by direct cell contact, or by cellular extensions (e.g. filopodia) [372]. Table 2 compares two systems where evidence exists about the molecular basis of the inhibitory agent: the spacing of heterocysts (nitrogen-fixing cells) in *Anabaena* (a cyanobacterium) and the spacing of bristles in *Drosophila*. The various bristle patterns that are supposedly created by inhibitory fields are depicted in figure 5.

The type of pattern studied by Wigglesworth can theoretically be generated de novo – rather than gradually through growth – by having naive cells randomly choose to adopt a particular state and immediately prevent their neighbors from doing so [114]. When the array becomes saturated with inhibitory fields, then the nearest-neighbor distances between the sites will range from 1.0 to 2.0 inhibitory-field radii. To produce a more uniform spacing of pattern elements, a 'competence wave' has been included in several models [116, 579, 659, 720, 995]. The term refers only to the *effect* of the wavefront – not its mode of travel, which could theoretically be via either a relay or a schedule. The idea, as illustrated in figure 4c, is that a cell is only 'competent' [925] to become black when it is within the wavefront. Upon becoming black, the cell immediately establishes an inhibitory field, so that the next cell that can become black will lie just outside the field. In the final pattern all of the nearest-neighbor distances would equal 1.0 inhibitory-field radius.

Inhibitory fields have been invoked not only to account for the patterning of heterocysts [596, 967, 968] and insect bristles [114, 115, 369, 370, 425, 487, 720, 831] as discussed above, but also insect neuroblasts [178, 180] (where laser-ablation of an incipient neuroblast leads to its replacement by an adjacent ectodermal cell) and pheromone glands (whose pores develop only at the vertices of polygonal epidermal cells) [813]; axolotl dermal glands (where nearest-neighbor distance is proportional to gland size) [385]; shark scales [718]; shark [717], reptile [657, 658], and mammal [89] teeth; sheep hairs [114, 117]; and leaf stomata [84, 85, 471–473, 752, 753].

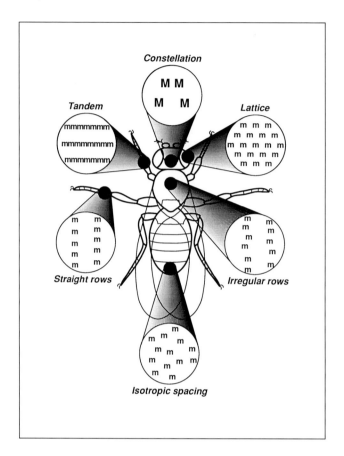

Fig. 5. Drosophila bristle patterns, supposedly created by inhibitory fields. On the fly surface there are roughly 5,000 bristles, a few dozen of which are much larger than the rest. These 'macrochaetes' (M's) function as extended mechanoreceptors (like cat whiskers) and tend to occupy highly stereotyped aperiodic ('constellation') patterns. The smaller 'microchaetes' (m's) tend to be evenly spaced and aligned to varying degrees. Most bristles on the dorsal abdomen are 'isotropically' spaced (i.e. no axial alignments; as in the insect studied by Wigglesworth [963]), while those on the thorax and legs are aligned in uniaxial rows, and those in the compound eye occupy vertices in a triaxial lattice of ommatidia. Bristles can be so close that they actually touch (the 'tandem' array). Based upon various lines of evidence (including mutant phenotypes), all bristle cells are thought to possess inhibitory fields, with the region-specific differences in bristle arrangements being due to temporal and spatial restrictions on exactly which epidermal cells are 'competent' to become bristle cells [365, 788]. From Held [369].

Models that include a competence wave (either linear or annular) in conjunction with inhibitory fields can produce amazingly precise patterns, including hexagonal lattices (bird feathers [158, 159, 196, 524], fly ommatidia [369], and, possibly, vertebrate photoreceptors [483, 965]) and arithmetic spirals (plant leaf primordia [593, 721, 768, 818, 883, 884, 936, 1016]) (fig. 6). The inhibitory-field/competence-wave mechanism thus provides one plausible solution to the 50-star portion of the 'American Flag Problem' posed in the Introduction.

Specific Inhibitor Model

Hierarchies of qualitatively different inhibitors could create an orderly series of different cell states [137–140, 580] – an idea first proposed by Rose [730–732]. He argued that cells having state 'A' could produce inhibitor 'A', which would diffuse locally, allowing cells outside the inhibitory field to assume state 'B' (the next state in the hierarchy), produce inhibitor 'B', and the process would thus continue until the tissue becomes partitioned into a series of stripes (A, B, etc.) whose relative widths could be controlled by the diffusion parameters of the inhibitor molecules. Since this model could easily produce alternating red and white bands, it shows how the 13-stripe portion of the 'American Flag Problem' could be solved without a coordinate system. Embellishing upon this simple model, Green and Cooke [321] have proposed a 'Serial Diversion' Model, which, interestingly, can scale a pattern in proportion to its size. Instead of inhibitory fields, they use a reaction-diffusion mechanism. A competence wave enables cells to make a succession of different activator and inhibitor molecules (cf. Meinhardt [573]), which diffuse to the boundaries of the tissue. Because their diffusion is confined, their concentrations will increase if the area decreases, allowing them to reach the 'diversion threshold' for the next activator and inhibitor at an earlier time, hence causing the competence wave to compress the entire pattern to fit the area. Bard [33] had earlier proposed a similar hybrid mechanism to explain how parallel stripes develop in zebras.

Specific Activator Model

The opposite of a Specific Inhibitor Model would be a 'Specific Activator Model', and this type of mechanism was advocated by Weiss [952]. He imagined that the confrontation of two tissues A and B could cause reactions at their interface, leading to the creation of an intermediate layer C, which could then react with both A and B, thus continuing to stratify the tissue ad libitum. Similar arguments for sequential interfacial inductions

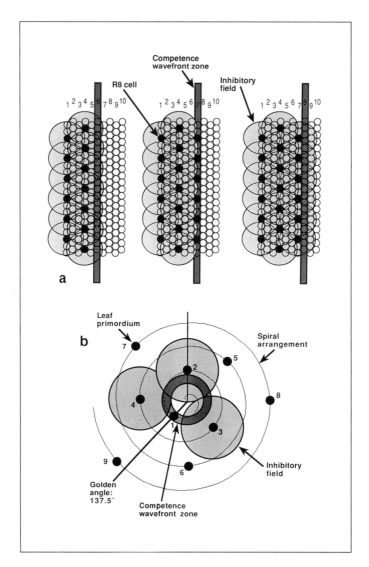

Fig. 6. Precise patterns, presumably created by inhibitory fields plus competence waves. Surprisingly, researchers working with geometrically different patterns in organisms as different as flies, birds, and plants have independently devised similar models to explain the precise patterning of ommatidia, feathers, and leaves respectively. *a* Cellular interactions which supposedly lead to the development of a hexagonal lattice of 'R8' photoreceptor cells in the *Drosophila* eye [27]. A 'morphogenetic wavefront' [714] (dark vertical bar) sweeps across the array, endowing cells with the 'competence' to become R8 cells. Cell columns are numbered so as to provide reference landmarks. Only a small

have been advanced to account for progressive differentiation in sea urchins [161] and *Xenopus* [557, 800, 802], and for intercalations of cell states between stripes of segment-polarity gene expression in *Drosophila* [377, 393, 492, 554, 555]. Aspects of Weiss's model also resemble elements of Meinhardt and Gierer's scheme for 'the lateral activation of mutually exclusive states' [573, 580].

---

portion of the eye rudiment is diagrammed. Three successive stages (left to right) in the progress of the wavefront are illustrated. Though an uncommitted (white) cell must be in the competence wavefront zone in order to become an R8 (black) cell, its presence there is not sufficient. The cell must also reside outside the inhibitory fields (shaded ovals) of pre-existing R8 cells. Cells that satisfy both criteria become R8 cells and establish new inhibitory fields. Thus, a hexagonal lattice forms because each new row of R8 cells arises in the interstices of the inhibitory fields of the previous R8 row (after Held [369]). The sites in the original (leftmost) column, which function here as a 'seed crystal', could conceivably have been established by an earlier perpendicular wave traveling along that column (as in fig. 4c). This model was first proposed as an explanation for the development of hexagonal patterns of bird feathers, which also develop in a wavelike progression within discrete epidermal 'tracts' [196]. In other types of models for hexagonal patterning, the competence wave has been retained, but inhibitory fields have been replaced by either a reaction-diffusion [623] or a physical-force [351, 662, 681] mechanism. A competence-wave/chemotaxis strategy has actually been used in vitro to coax chemotactic bacteria to self-assemble into a lattice [82]. *b* Cellular interactions supposedly occurring in a plant apical meristem. The leaves of many plants originate in a spiral pattern. Along the cylindrical stem created by the dome-shaped meristem, the spiral becomes a helix (not shown). Because the positions of new leaf primordia (small filled circles) can be altered by surgically separating their presumptive sites from older adjacent primordia, it seems that older primordia possess inhibitory fields (shaded circles) [818, 936]. Furthermore, because primordia arise at a constant distance from the apex, there appears to be a competence zone (dark annulus) analogous to the morphogenetic wavefront in the *Drosophila* eye. However, in this case, successive ranks of cells move through the (stationary) wavefront zone (by centrifugal displacement away from the mitotically active apex), rather than the other way around, and the zone is circular instead of linear. Interestingly, mathematical simulations have shown that inhibitory fields, centrifugal growth, and a competence zone are sufficient to generate spiral arrangements – even without a 'seed crystal' to start the pattern [593, 721, 883]. Moreover, the simulations explain why successive primordia in so many plant species arise at the 'golden angle' of 137.5°. The explanation is purely steric, as depicted here: given the relative overlaps of the wavefront zone with the inhibitory fields of the three most recent primordia (No. 2, 3, and 4), the next primordium (No. 1) must arise at approximately this angle relative to its immediate predecessor (after Richter [721] and Wardlaw [936]). Hexagonal lattices and arithmetic spirals are also found in the realm of animal behavior, i.e. honeycombs and orb webs, and there too, only a few cue-driven behaviors seem to be involved [51, 194, 195, 682, 716, 922, 985].

Chapter 4:
Darwinian Mechanisms

'Sibling rivalry', as exemplified by the contests between *Anabaena* proheterocysts or *Drosophila* bristle cells (table 2), is merely one form of intercellular competition – a phenomenon observed commonly in embryos [429, 588, 810]. In 1881, 22 years after Darwin's 'Origin of Species', a book entitled '*Der Kampf der Theile im Organismus*' [738] ('The Struggle of Parts within the Organism') was published which asserted that embryos develop in a manner analogous to how species evolve – by selection among competing variants (in this case, cells) for an adaptive outcome (a functional anatomy) [653, 655]. The author was Wilhelm Roux, a student of Haeckel's and the founder of *Entwicklungsmechanik* (the science of developmental mechanics) [30, 602, 739]. Since then, much evidence has been adduced for selective strategies in development, and Gerald Edelman's recent book 'Neural Darwinism' [202] has rekindled interest in such mechanisms, especially as they pertain to the nervous system.

Cell Death Models

One of the most perplexing events in development is cell death [763]. Why should an embryo invest its precious energy in creating new cells, only to destroy them? Aside from serving a sculpting function during morphogenesis and metamorphosis [378, 530, 763], there appear to be several pattern-related reasons for why cells die:

(1) *Being born into a 'pruned' lineage tree.* Studies of cell death mutants in *C. elegans* have led to the conclusion that certain cells are autonomously 'programmed' to die [213]. Given the evolutionary conservatism of many of the lineage motifs [841, 842], such retroactive alterations may be genetically simpler than the redesigning of entire pathways. Comparable cell deaths are observed among bristle and scale lineages in insects [485], as well as among identified neuroblasts [531], perhaps for the same reason.

(2) *Residing in a sagging gradient.* According to hybrid 'gradient/reaction-diffusion' models, a constant number of strucures can be generated within a field of cells if the size of the array is tightly controlled [134, 136, 138, 1008], and that control could be mediated by the slope of the gradient. Lawrence [490] has suggested that when the slope is too shallow, one way to steepen it is to contract the cell array by interspersed cell death, as does indeed occur in mutant *Drosophila* embryos whose segments are wider than normal [546] (cf. Klingensmith et al. [465]).

(3) *Waiting too long to select a state.* On average, 2 or 3 cells per ommatidium die during the construction of the *Drosophila* retina, and mutations are known which simultaneously rescue the cells and distort the lattice [987]. The implication is that the extra cells normally die because they 'fail to establish contacts appropriate to a specific cell type', and that their elimination is necessary to 'tighten' the lattice [91]. The existence of such a 'clean-up' mechanism may explain why so many genes can mutate to give reduced-eye phenotypes [246, 523]: slight timing errors [92] could easily lead to massive cell death.

(4) *Matching or mismatching.* There are two major vertebrate systems that employ the bipartite strategy of (1) creating an excess of cells, and then (2) selecting a subset thereof: the immune system and the nervous system. In the immune system, antibody diversity is generated by gene rearrangements within lymphocyte progenitors [197]. Those lymphocytes bearing 'anti-self' antibodies are then eliminated (negative selection) [742, 772, 923], while those bearing 'anti-invader' antibodies proliferate during each infection cycle (positive selection) [3, 974]. In the nervous system, excess neurons are produced in virtually every region, the competition for targets is intense, and many neurons die because they fail to find their targets [148, 654, 702]. For complex neural networks, the selective strategy is advantageous insofar as (1) it ensures numerical parity of the matching populations [112, 435, 436, 973], and (2) it can fine-tune the interconnections based upon functional criteria [74, 430, 701, 702]. The latter point is one of the central tenets in Edelman's Neural Darwinism Theory.

Given these examples where cell death is used as a patterning tool, the cell-death mechanism depicted in figure 7a is plausible, though no actual *periodic* patterns have been shown to originate in this manner thus far.

Most deaths in development are suicides rather than murders [113, 771]. A notable exception is the engulfment and murder of the *C. elegans* gonadal linker cell by the E.lp and E.rp killer cells. Laser ablation of the parent cell before E.lp and E.rp are born rescues the linker cell [213, 864].

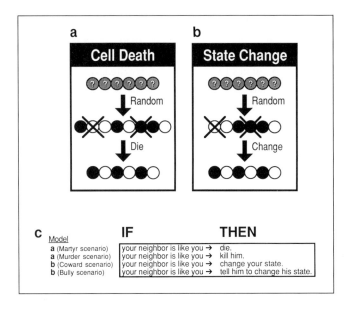

Fig. 7. Darwinian mechanisms. In Darwinian models, cells acquire states (randomly or in some other manner) and the crude pattern is then 'fine-tuned' by either cell death or state change. Both of the following models have a cellular automata format, but they lack a determination wave. *a* Cell Death Model. Wherever two alike cells are adjacent, one dies – either by murder or suicide. The choice of which neighbor dies could be stochastic. (The 'die' rule is reminiscent of Conway's game of 'Life' [254]). *b* State Change Model. Wherever two alike cells are adjacent, one changes its state. Determination of which cell changes could be stochastic. More than one round of state changes might be required to erase all matches. *c* Conditional 'IF/THEN' rules implicit in the models depicted above.

State-Change Models

There is evidence that morphallactic regulation may involve a 'ripple effect', wherein cells throughout a surgically reduced organ change their states in response to changes in the states of their neighbors until the original coordinate system is compressed to span a smaller cellular array [242]. In his 'Adaptive Antichaos' Model, Kauffman has devised rules for how the states of a cell's (or gene's) neighbors (or regulators) can influence its own choice of states; and by setting Darwinian criteria for certain types of mismatches, he has shown that Boolean networks of such elements can develop stable, orderly patterns [447]. A similar 'living network' scenario (involving sequential influences among neighboring elements) is the centerpiece of Edelman's 'Topobiology' Theory [203–205], except in that case, the elements are *cells* on the one hand and inanimate *extracellular substrates* on the other. These two entities communicate via 'CAMs' [198, 199, 201] (*cell* adhesion molecules) on the former and 'SAMs' (*substrate* adhesion molecules) on the latter. In response to position-specific SAM signals, a cell can: (1) change its CAMs; (2) cause the SAMs to change; or (3) react by choosing any response in the *accessible* subset of its genetic or behavioral repertoire (accessibility is supposedly under CAM control), including moving (steered by the CAMs [200]) to another SAM environment, where a new dialog can begin [271]. Kauffman's and Edelman's models are ornate versions of 'Turing machines' [394, 906] and, as such, serve as testable incarnations of the 'Cybernetic Metaphor' for embryonic development (see chap. 7).

A state-change mechanism, such as the one diagrammed in figure 7b, may be involved in the patterning of lepidopteran wing scales. In several genera, the scales are arranged in parallel anteroposterior rows [1015], and within each row there are two types of scales: cover ('*c*') and basal ('*b*') scales which alternate (*cbcbcbcbcb* ...) [1014]. In *Pieris,* each scale row develops from a single file of scale precursor cells (with no ordinary epidermal cells intervening within the file), which are also of two types: wide ('*w*') and narrow ('*n*'). However, the alternation of the *w* and *n* precursors is less perfect: 35% of the [*wn* + *wwn* + *wnn*] sequences have homotypic pairs [*wwn* + *wnn*], whereas only 4% of the total [*cb* + *ccb* + *cbb*] sequences include them [*ccb* + *cbb*]. Cell death and cell movements have been ruled out as explanations for the fine-tuning process [1014], leaving state changes as a possibility (e.g. *wwn* → *nwn* → *bcb,* and *wnn* → *wnw* → *cbc*).

Chapter 5:
Rearrangement Mechanisms

Tissue movements figure prominently in the early development of most metazoans, and individual cells frequently rearrange or migrate [247, 452, 899]. A cell's state of determination or differentiation can theoretically cause it to assume a specific position relative to other cells (by propelling it in a particular direction, or by stimulating it to move until a certain condition is met) [947]. Clear-cut evidence for such a mechanism has recently been found for identified motoneurons in embryonic zebrafish: 'after they were transplanted to new positions, the somata of many primary motoneurons moved back to their original positions' [209].

Adhesion Models

Many viruses [10, 290, 516] and subcellular structures [10, 329, 412, 461, 921, 952, 953] are capable of 'self-assembly'. As exemplified by the Watson-Crick pairing of nucleotide bases, the process usually relies upon a 'jigsaw puzzle' fitting together of complementary binding sites on the participating monomers, which yields a configuration of minimum free energy. The idea of a self-assembling *supra*cellular neuro-architecture, based upon the same jigsaw-puzzle metaphor, was crafted by Roger Sperry in a series of papers in the 1940s (it is traceable to Langley, 1895) [481, 700] and cogently summarized in 1963. It was Sperry's [808] conjecture that

... the cells and fibers of the brain and cord must carry some kind of individual identification tags, presumably cytochemical in nature, by which they are distinguished one from another almost, in many regions, to the level of the single neuron; and further, that the growing fibers are extremely particular when it comes to establishing synaptic connections, each axon linking only with certain neurons to which it becomes selectively attached by specific chemical affinities.

A mechanism of this kind operates in the central nervous systems of insects, where different axon pathways express different surface antigens [43, 301]. Based upon ablation experiments, surface-labeling studies using monoclonal antibodies, and other lines of evidence, Goodman et al. [302] proposed a 'Labelled Pathways Model', which contends that neuronal pathfinding in the insect CNS is accomplished by qualitatively different labels that are 'read' by the growth cones of migrating axons.

Sperry's Chemoaffinity Model [314] has been eliminated as a viable explanation for the afferent connections between retinal axons and tectal neurons in the amphibian visual system. One item of counterevidence [702] is the ability of the 'projection pattern' (i.e. the pattern of retinotectal linkages) to be compressed or expanded to accommodate the relative sizes of the two organs when they are altered surgically [192, 193]. Such adjustments should be impossible if specific retinal axons are programmed to bind to specific tectal cells. In its place, many other models have been proposed, most focusing on single behaviors in the neural repertoire [128, 238, 239, 702]. An exception is the Synthetic Model of Fraser and Hunt [236, 237, 239], which has the virtues of being able to explain (1) minority as well as majority results from a number of experimental regimens, and (2) the formation of 'ocular dominance stripes'. These peculiar stripes develop when axons from an extra transplanted eye try to project onto the same tectal surface as axons from the in situ eye [128]. The Synthetic Model invokes three forces: (1) an adhesive 'C' force between retinal and tectal cells regardless of origin; (2) a position-specific retinotectal adhesive force, which is graded along the AP and DV axes; and (3) a repulsive 'R' force between retinal fibers (with the relative strengths assumed to be 'C > R > DV > AP' in *Xenopus*). If the R force depends upon correlated electrical activity in neighboring axons (i.e. 'fibers that fire together synapse together' [237]), then ocular dominance columns (which are uniformly about 200 μm wide [128]) are explicable because the 'xenophobic hatred' of the fibers exceeds their 'domestic distaste'. Neighboring retinal cells do indeed fire synchronously in mammals [582]. In the mammalian CNS, analogous *naturally occurring* stripes characterize intrinsic and descending projections of the neocortex, projections to the cerebellum, binocular retinal projections to the superior colliculus, and afferent projections from the lateral geniculate nucleus to the visual cortex [128, 131, 401], and activity-dependent rules may likewise explain the periodicities of these afferent segregations [130, 780, 1019]. Interestingly, these rules are formally the neural equivalents of the 'local-activation-lateral-inhibition' rules in reaction-diffusion models for fingerprints and zebra stripes [589, 660, 872]. In an article entitled 'Thinking about the brain' [150] Crick once mused that 'there is something in embryology that likes stripes', and the prevalence of such mechanisms in development may explain why.

Axial gradients of potential adhesion molecules have accordingly been found in the retinotectal systems of birds [902] and rats [129]. Elsewhere, adhesion gradients have been implicated in lepidopteran wings (proximo-distal axis) [631, 632], *Drosophila* wings (radial, as a function of distance from the wing margin) [725], and cotton bug tergites (anterior-posterior axis) [648], where they may provide haptotactic guidance cues [437, 900] for migrating axons [630], bristle cells [725], or myocytes [102, 972], respectively. Along the proximodistal axis of the amphibian limb, graded adhesive differences have also been demonstrated, which may play a role in the communication of positional information [634], and an adhesivity gradient likewise appears to guide the migration of the amphibian pro-nephric duct [1017, 1018].

'Sorting out' is the process whereby cells of different types segregate after being mixed and aggregated in vitro. Many cell mixtures behave in this manner [225, 226, 898, 951], demonstrating that the state of a cell can indeed cause its position, at least under these artificial conditions. Sorting out per se does not prove an adhesive mechanism since differential che-motaxis – e.g. in response to an oxygen gradient caused by respiration of the aggregate – could also cause sorting out [541, 898]. Indeed, chemotactic sorting has been demonstrated for prestalk and prespore cells in *Dictyostelium* [846, 874]. Classic experiments by Townes and Holtfreter showed that amphibian cells from different germ layers can segregate into concentric layers that mimic their normal stratification in vivo [897] – suggesting that a 'self-assembly' strategy might be used during normal development (but in amphibia, neither gastrulation nor neurulation proceeds via cell sorting). Steinberg [819] conducted numerous similar experiments with a variety of different tissues and discovered an 'Adhesive Hierarchy Rule': if tissue A sorts to the inside of an aggregate when mixed with tissue B, and B likewise sorts inside C, then A will sort inside C. Based upon this rule and other evidence, he proposed a 'Differential Adhesion Model' [820]. The model assumes that cell types vary in the degree of their adhesivity, and the 'stickier' the cell type (A > B > C) the more it will tend to form a tight and homogeneous clump [821, 823]. Investigations of neuronal pathfinding in the insect peripheral nervous system led Berlot and Goodman [57, 300] to propose a similar 'Adhesive Hierarchy Model':

Our results suggest that the filopodia of the pioneer growth cones in the antenna and limb buds of the grasshopper embryo express an adhesive hierarchy, whereby the surfaces of neurons are preferred over the surfaces of the epithelial cells. (1) Given only the epithelium, the growth cones extend proximally along its surface, appearing to follow an epithelial adhesive gradient. (2) Given a choice in the periphery, however, of neurons versus epithelium, the filopodia preferentially adhere to the neuronal surfaces and thus guide the growth cones onto these neuronal cell bodies and axons. (3) Given a choice in the CNS of different axon bundles, certain neuronal surfaces appear to rank higher in the adhesive hierarchy than others; they invariably choose a particular axon bundle on which to extend, similar to the observation of selective fasciculation by central neurons that led to the labeled pathways hypothesis [57].

An 'Adhesivity/Determination-Wave' hybrid model has been proposed for somite formation in chick embryos based upon observed increases in the adhesivity of mesodermal cells before and after somite formation [35, 53, 108]. The idea is that paraxial mesoderm cells (perhaps gated into groups by the cell cycle [455]) increase their adhesivity when a competence wave reaches them, and clumps of them (nascent somites) consequently pinch off sequentially from the segmental plate.

Both homophilic and heterophilic adhesion molecules have been discovered [41, 176, 198, 282]. The patterning scheme in figure 8a employs heterophilic molecules at the poles of each cell to assemble an alternating chain of two cell types. Head-to-tail associations of this kind are critical for differentiation of *Myxococcus* bacteria [458]. A dramatic illustration of self-assembly in vitro is the spontaneous alignment of human fibroblasts [216] or keratinocytes [320] into 'fingerprint-like' [154] patterns containing loops, whorls, and triradii (cf. Nübler-Jung [647] and Seul et al. [777]).

Repulsion Models

The movements of chromatophore cells emigrating from amphibian neural crest explants in vitro (and the spreading behavior of such cells from transplants in vivo) led Twitty and his collaborators [908, 910, 911] (but cf. Erickson and Oliver [217]) to conclude that:

... the cells move in response to a mutual stimulation, or 'repulsion', probably mediated through the action of diffusible substances involved in or produced by the metabolism of the cells (themselves) ... with the cells gradually spreading peripherally until they are spaced beyond the effective range of such mutual influences [908].

One of their experiments was especially informative [912]. Individual chromatophores were sucked into a capillary tube (full of coelomic fluid) and observed for several days. When there was only a single cell in the tube, it moved little. When two cells were placed together, they moved in opposite directions; and when there were three cells, they spread out until their spacing was uniform – all of which are behaviors consistent with Twitty's 'Mutual Repulsion Hypothesis' (fig. 8b).

In certain urodele amphibians (e.g. *Taricha torosus*) the pigment cells secondarily (i.e. following their initially uniform 'primary' distribution) aggregate into bands on the dorsal flank of the tadpole [904, 905]. The aggregation behavior also occurs in vitro and is attributable to direct cell contacts – via filopodia – which pull neighboring cells together [909]. In the frog *Bombina,* a gossamer network of cruciform filopodia develops from one class of (adepidermal) melanophores. Ellinger described the stages in its formation:

> To this stage [6 days postfertilization], the melanophore distribution and orientation of cell extensions appeared to be random. This arrangement was altered markedly during the seventh day of development. A pattern became established in which the epidermal melanophore extensions frequently left the cell body at right angles to each other ... As differentiation proceeded, a 'grid' of interwoven epidermal melanophore processes became apparent within the integument ... [211].

Based upon the behavior of the filopodia at their intersections, Ellinger argued that the filopodial interactions *cause* the rearrangement of the lattice.

A similar causal role has been ascribed to filopodia on the scale cells of moth wings [633]. The scale cells in *Manduca sexta* are initially randomly arranged. They gradually align and become evenly spaced by rearranging, during which time the distances between neighboring scale cells are spanned by axially oriented filopodial extensions ('epidermal feet'). Whether the filopodia push or pull or simply serve as 'phone lines' for the communication of intercellular navigational signals is not known. In *Drosophila melanogaster,* mutations in several unlinked genes cause the misorientation of bristle cells, and the intervals that separate neighboring bristles (which are uniform in wild-type flies) are correlated with the relative orientations of the cells (bristles facing one another are far apart, while those facing away from one another are close together), suggesting a repulsive interaction via unidirectional filopodia [371]. (Bristle and scale cells are thought to be homologous [485, 917, 960].)

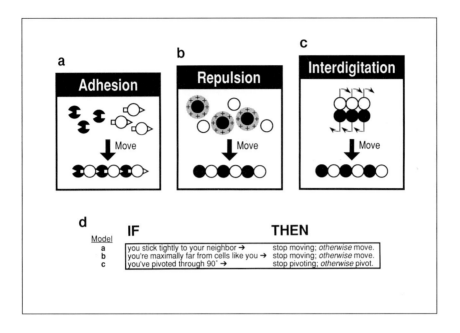

Fig. 8. Rearrangement mechanisms. Rearrangement models assume that the states of cells cause them to relocate. The starting configuration could be the outcome of an earlier round of patterning (via a different mechanism) or simply random. Whether intercellular communication is required depends upon the mode of rearrangement. *a* Adhesion Model. Cells have complementary binding sites on their surfaces, represented here as jigsaw-puzzle protrusions and indentations. The cells move until their binding sites are occupied. Depending upon the dimensions (1, 2, or 3) of the final array, its geometry (striped, checkerboard, etc.), and the numbers of cells in each homotypic domain, different types of binding (homophilic vs. heterophilic, quantitative vs. qualitative, etc.) might be needed. *b* Repulsion Model. One of the two cell types (black) is assumed to repel cells of its own kind. The mechanism of repulsion could be electrical (as depicted here, though galvanotaxis seems to play little role in development [282]), mechanical (via filopodial extensions like growth cones [372]), or chemical (via chemorepellent molecules [912]). The model is more easily visualized for a two-dimensional array (where cells can jostle within a fluid monolayer) than for the one-dimensional situation depicted here (where cells must move out of alignment in order to rearrange, and a separate mechanism, not shown, must be invoked to realign them). *c* Interdigitation Model. Different cell types, that are originally segregated into separate files, merge into a single file. Here the interdigitation is accomplished by choreographed rotations of heterotypic pairs. *d* Conditional 'IF/THEN' rules implicit in the models depicted above.

Growth cones of navigating axons (an extreme version of a filopo-
dium) can also respond to inhibitory (repulsive?) signals from other axons
[164, 454]. The possibility that attraction and repulsion utilize the same
signaling mechanism is suggested by the phenotypes of *egl* mutants in *C.
elegans,* where a normally attractive target becomes repellent [838].

Interdigitation Models

In addition to being able to migrate individually (e.g. neural crest
cells) and jostle within a moving monolayer (e.g. mesodermal cells during
archenteron invagination) [247, 452, 899], cells can apparently also
'dance', i.e. perform choreographed maneuvers with one or more partners.
For example, as preclusters of photoreceptors emerge from the morphogenetic furrow in the developing *Drosophila* retina, the cluster cells are
arranged in an arc-shaped single file, but shortly thereafter they close ranks
and form a rosette [986]. Also in the same tissue layer, the four cells of each
bristle organ align themselves before enveloping one another concentri-
cally [91]. Neither of these rearrangements actually produces the larger
periodic patterns to which the cell groups belong. However, there is evidence that some periodic patterns do form via 'minuets' wherein the cellular partners pivot and interdigitate (fig. 8c). Thus, there is a row of bristles (row 8) on the tarsus of Hawaiian *Drosophila* species where the bristles
tandemly alternate with 'bracts' in a single file [367]. Bracts are noninner-
vated cuticular protuberances, which develop from epidermal cells that
have been induced by the bristle-cell complex [370]. Based upon what is
known about tarsal cell lineage in *D. melanogaster* [366, 493], it is likely
that the bristle and bract cells originate lateral to each other and subsequently interdigitate, perhaps by the pivoting of bristle-bract cell pairs.
Pivoting may likewise occur in the wings of certain lepidopteran species,
where the precursor cells for the basal scales apparently insert themselves
(at regular intervals) into homogeneous rows of cover-scale precursor cells
[862, 1014]. Quarter pirouettes (90° rotations) have been observed in two
other systems:
    (1) The hexagonal array of photoreceptors in *D. melanogaster*. Photo-
receptor clusters in the ventral half of the eye swivel 90° clockwise and
those in the dorsal half swivel 90° counterclockwise, so that a new (some-
what zigzag) plane of mirror symmetry is established at the equator
[893].

(2) Blocks of myotomal cells in *Xenopus*. Somites swivel 90° clockwise on the right side of the body and 90° counterclockwise on the left side [346]. Comparable rotations occur in chick embryos, though in that case they happen after the somites have segregated from the paraxial mesoderm [52]. (In *Xenopus* they occur during segregation.)

In neither the *Drosophila* eye nor the vertebrate spine do the rotations *cause* the periodicity of the patterns that they modify. The rotations may serve an optical function in *Drosophila*. (Other insect eyes are highly stratified along the dorsoventral axis [296].) In *Xenopus* and chick embryos, the purpose of the pirouettes is less apparent. (Somites in most vertebrates do not rotate [135].)

Chemotaxis Models

The chemotaxis models of Oster, Murray, and others [620, 622, 663] behave in a manner that resembles the aggregation of *Dictyostelium* cells [172], except that the diffusible signal is neither pulsed nor relayed: all cells are assumed to emit a chemoattractant continuously. If the cells are initially scattered within a linear stripe, then the stripe will automatically dissolve into a series of clusters, each of which becomes more attractive as it forms, depleting the surrounding areas of motile cells. Such models are formally equivalent to reaction-diffusion schemes insofar as they also rely upon local autocatalysis and lateral inhibition [663]. A developing system where a narrow stripe actually does fragment into separate islands of cells (perhaps by this sort of mechanism) is the supraorbital lateral line primordium of *Xenopus* [983, 984], and a peculiar dispersal of tissue fragments has been described for pigment cell clusters in the fish *Blennius pholis* [901]. In accordance with the predictions of this sort of model, a suspension of chemotactic bacteria can indeed coalesce into a precise lattice of clumps if the stimulus that initiates the chemotaxis spreads centrifugally (like a competence wave) through the array [82].

Chapter 6:
Cell-Lineage Mechanisms

Until about 1900 it was customary to characterize embryos as 'mosaic' or 'regulative', based upon whether isolated blastomeres form only their normal portion of the anatomy (mosaic) or a larger portion (regulative) [583, 602, 998]. Gradually, the value of the distinction waned, as it became clear that the outcomes of such experiments depended critically upon the stage at which they were performed [160, 408, 945]. However, a version of the dichotomy has persisted until only recently – namely, the notion that embryos with 'determinate cleavage' (precise sequences of cleavage planes) assign cellular fates by 'cytoplasmic determinants' (substances that are asymmetrically partitioned to daughter blastomeres), whereas those with indeterminate cleavage assign fates via cellular interactions [160, 161, 602, 975]. In the last few years, the conceptual wall separating these two categories has begun to crumble, thanks mainly to nematodes and sea urchins, where determinate cleavage is combined with regulative ability. Ablations or transplantations in these species can alter cell fates, thus implicating cellular interactions (rather than the rigid cellular pedigrees) as the causative agents in the assignment of those fates [161, 459] (cf. Dohle [183]). Citing such cases, which are heretical exceptions to the old stereotypes, Davidson has devised a more pluralistic classification scheme for incorporating the role of cell lineage in embryonic development [163], plus a set of basic models for the spatial regulation of histospecific genes [162] (cf. Holliday [388]).

In the wake of the old paradigm's demise, the cautionary lesson is that a ritualized cell lineage per se does not prove that cells are assigned their fates by means of that lineage [161, 459, 779, 829, 1000]: 'a determinate lineage may precisely locate cells, which then go on to adopt highly predictable fates because of extracellular cues' [244]. Only when relative cell positions are experimentally altered can the mechanism be ascertained. By contrast, it is easy to *disprove* a cell-lineage mechanism. For instance, by using a pigment mutation to genetically mark individual cells in the developing retina of *Drosophila* and then charting the locations of their descendants in the photoreceptor array, Ready et al. [714] were able to reject the hypothesis that each ommatidium is descended from a single mother cell. (They also showed that the equator of the eye is not established clonally.) Indeed, as a rule, organisms whose embryos manifest indeterminate cleavage do not construct their organs as clonal modules [151, 861].

Exceptions to this rule include certain miniature organs, e.g. insect sensilla [485] and leaf stomata [684, 816, 817]. These structures do develop according to strict pedigrees, though here too, some of the lineages may not be *causal* in assigning fates. Thus, each mechanosensory bristle in *Drosophila* is constructed from four cells – a shaft cell, a socket cell, a neuron, and a thecogen [451, 688] – all of which are descended from a single mother cell via two differentiative divisions [355]. The shaft and socket cells are sisters, as are the neuron and thecogen [91, 355], and the orientations of the mitotic spindles during these divisions are highly reproducible from one individual to the next [400]. Such rigid cell lineages are theoretically sufficient to assign fates, but in fact they can be bypassed: along the wing margin, where hundreds of bristle cells jostle into alignment, sister cells can contribute to separate bristles and adopt the same fate (e.g. socket cells) [355]. It is their relative positions that appear to determine their fates – a supposition which is supported by the changes in cell fate that accompany mutational alterations in bristle cell positions [29, 498].

Three types of possible lineage mechanisms are discussed below, along with various developing systems whose cell lineages conform to one or another of the schemes. Unfortunately, in most cases the critical experiments have not been performed to ascertain whether the lineage plays a causal role.

The Quantal Mitosis Model

The term 'quantal cell cycle' was introduced by Holtzer [390–392] to designate an asymmetric cell division in which the parent cell differs from at least one of its daughters, in contradistinction to a 'proliferative' cell cycle, where parent and daughters are identical. An example of a quantal cycle was mentioned for *Anabaena* in chapter 3 (table 2): every mitosis along the filament is asymmetric, yielding a small daughter that is capable of becoming a heterocyst, and a large daughter that is not [594, 967]. The asymmetric mitoses in this case, however, are not solely responsible for the

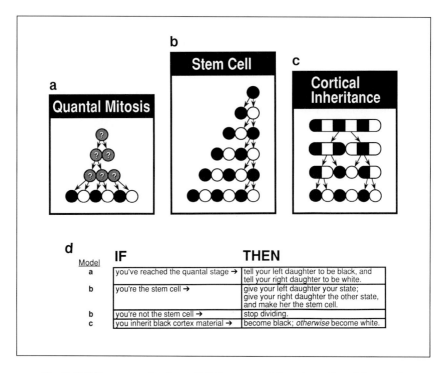

Fig. 9. Cell-lineage mechanisms. Cell-lineage models assign cell positions and states via strict pedigree rules, with no involvement of intercellular communication. *a* Quantal Mitosis Model. Uncommitted cells undergo proliferative mitoses until they reach a critical stage, at which time they undergo a 'quantal' mitosis. In each case, the daughter on the left is instructed to become black, the other one white. *b* Stem Cell Model. A single stem cell divides repeatedly, changing its state every mitosis so that it leaves behind a chain of alternately black and white daughter cells. *c* Cortical Inheritance Model. Stripes of regulatory molecules are established in the cortex of the progenitor cell, and the cortical molecules cause descendant cells to adopt particular states. *d* Conditional 'IF/THEN' rules implicit in the models depicted above.

regular spacing of heterocysts, since inhibitory fields play a crucial role. Other periodic patterns where quantal mitoses are involved in the spacing of pattern elements include:

(1) The alternating pattern of hair cells ('trichoblasts') and ordinary epidermal cells in the root epidermis of various plant species [155, 184], including the monocotyledon *Phleum pratense.* 'The last division [of the precursor cells] in a plane perpendicular to the longitudinal axis of cell

polarity gives rise to two daughter cells. The division is unequal, a smaller cell with denser cytoplasm being cut off toward the apical (distal) end, and a larger cell at the basal (proximal) end. The smaller cell (trichoblast) remains strongly meristematic and soon forms ... a root hair' [62].

(2) The spacing pattern of gonidial precursor cells in *Volvox* (a colonial green alga [462]). At the sixth cleavage in *Volvox carteri,* 'all 16 (or under suboptimum conditions only a portion) of the cells, which were derived from the two anterior tiers of the 16-cell embryo, divide unequally to yield a small-anterior/large-posterior pair of sister cells. The larger member of each pair becomes a gonidial initial; the smaller, a somatic initial' [463]. Surgical experiments with a related species, *Volvox obversus,* have shown that size alone is not the determining factor [705]. The distances between the gonidial initials are subsequently increased by further divisions of the intervening somatic cells and by several additional asymmetric mitoses of the gonidial cells themselves.

In both of the above examples, the cellular array constitutes a 'tessellation' [288, 330] pattern, since it fills space using a single type of tiling unit. Here, the unit is a clone containing one pattern element plus one or more 'background' cells (which are generated by the same compartmentalized mechanism that produces the pattern element itself). Other systems where pattern elements appear to be positioned by a clonal tesselating mechanism include: wing scales in certain butterflies [487, 948] and leaf stomata in certain plant species [84, 152, 471, 751, 753]. In the hypothetical illustration in figure 9a, every cell (at a definite time in development) undergoes a mitosis which is both asymmetric (the daughters are different colors) and polarized (black is on the left and white is on the right), resulting in a periodic pattern.

The Stem Cell Model

The stem-cell strategy is a variation on the quantal mitosis theme [1006]. In this case, only one cell undergoes asymmetric divisions, and it does so repeatedly (fig. 9b). Many organisms employ stem cells as pluripotent progenitors for a variety of terminally differentiated cell types [342, 534, 695]. In order for a stem-cell mechanism to directly generate a periodic pattern, it must create alternatingly different daughters. Such a process produces certain chains of cells in leech [779] and earthworm [849] embryos. Both of these annelids have determinate lineages in which 'bandlets' of ectodermal and mesodermal precursor cells arise from iterated mitoses of huge stem-cell 'teloblasts'. There are 5 teloblasts on each side of the midline, and 2 of them (N and Q) produce (n or q) bandlets within which every other 'blast cell' eventually manifests a different cell-lineage pattern. In the n bandlets of the leech, the alternate blast cells (n_s and n_f) also show a different staining intensity of a tracer dye when it is injected into the N teloblast, suggesting a difference in diffusibility through the cytoplasmic bridge to the parent teloblast when they are born [59, 60]. Bissen [59] incorporated this observation into a 'Flip-Flop Feedback Model', in which good diffusibility allows a blast-cell daughter to signal the teloblast to change its state (e.g. from 0 to 1) while at the next mitosis poor diffusibility would block the feedback signal and cause the teloblast to revert to its ground state, resulting in a flip-flop alternation of states: n_s, n_f, n_s, etc. Interestingly, segment number – which is extremely precise in leech species – is apparently not controlled by cell lineage [778] (cf. Pfannenstiel [683]).

An organism whose lineage has been analyzed completely is the nematode *C. elegans* [843, 865]. Kimble [459] has shown that even the most complicated lineage trees can be reduced to 2 elemental components – a 'stem cell' motif and a 'symmetrical mitosis' motif – that have been embellished in 5 possible ways: (1) a new switch in cell fate at some point in the pedigree; (2) a polarity reversal where fates are conserved but transposed; (3) a duplication of a cell near the beginning of a tree; (4) a duplication of a cell near the end of a tree; or (5) an iteration of one of the previous alterations.

L-Systems and Fractal Geometry

In many developing systems, cells 'self differentiate' [343, 408], automatically passing through a sequence of states in which each state change (of gene expression or cellular activity) is elicited by the previous state: state A → state B → state C, etc. [19, 95, 106,428, 533, 592, 811, 885, 945, 948]. By combining this notion of internally motivated state changes with a stem-cell style of meristematic growth, Aristid Lindenmayer invented a class of mathematical models that has since come to be called 'L-systems' [518–520, 522]. Like 'Turing machines' [174, 394, 906], L-systems postulate transition rules of the form 'IF you are in state *x*, THEN adopt state *y*'. By allowing more than one cell to be a stem cell, an L-system can cause the main filament to form branches that continue to grow and change states independently. (Because multiple growth points require parallel computations, such L-systems resemble cellular automata [886].) Furthermore, because the transition rules can lead to recursive cycles, the branches can undergo further branching on a smaller scale, resulting in 'fractal' patterns [40, 426, 550], which possess the property of 'self-similarity' (i.e. their structure remains the same at different levels of scale). Many biological patterns (e.g. trees, fern leaves, and blood-vessel networks) are fractals and can be simulated in this manner [520, 521, 656, 699, 740] (cf. De Reffye et al. [167]).

The Cortical Inheritance Model

In *C. elegans*, mutations in the gene *lin-17* cause asymmetric mitoses to become symmetric throughout various lineages [363, 844], implying that the assignment of cell fates may be controlled by 'switch genes' that are regulated in a cell-cycle-dependent manner [940]. Asymmetric mitoses are generally thought to result from qualitatively or quantitatively unequal allocations of gene-regulatory molecules to the two daughter cells [228, 363, 795, 797] (cf. Gober et al. [289]), though the control of mating-type interconversion in yeast (where each daughter inherits a different 'cassette' gene at a particular locus) shows that other mechanisms are also possible [375]. The regulatory molecules could either reside in the cytoplasm or the cortex [998]. Many examples of 'cytoplasmic determinants' are known [106, 549, 860]. By contrast, the eukaryotic cell membrane is usually viewed as a 'fluid mosaic' [281, 792], incapable of reliably partitioning regulatory signals because it cannot rigidly hold an array of molecules. However, epithelial cell polarities challenge this notion [247, 526, 528, 529]. The anisotropically pigmented cortices of many fertilized eggs (e.g. the dark animal hemisphere in *Xenopus*) are a vivid demonstration of the ability of cell surfaces to maintain large discrete domains, and there is evidence that they can stably maintain microdomains as well [329].

Aside from the remarkably periodic patterns of cilia in protistans [235, 804], there is at least one known case of a eukaryotic cell that generates a periodic pattern of molecules in its cortex. This is the *Drosophila* syncytial blastoderm, which is technically one cell despite its thousands of nuclei. It produces pair-rule stripes just beneath its cell membrane. The stripe widths and intervals do not depend upon nuclear spacing, since both remain uniform in haplodiploid mosaics where nuclear sizes vary [492, 863]. Conceivably, comparable scaffolds of molecules could be assembled within *any* cell (not just an egg), and future division planes could partition the cortex in a coordinated fashion (fig. 9c). In that case, the descendant array would constitute a cellularized version of the ancestor cell's cortex. Evolutionary reversals in the polarities of nematode lineages have been explained using this same rationale: the assumption is that the progenitor cell has at least three domains (A, B, and C) that can be segregated in different sequences (ABC → A/BC → A/B/C, or ABC → AB/C → A/B/C) by eccentric placement of successive division planes [841].

Chapter 7:
The Computer Metaphor in Developmental Biology

Watching time-lapse movies of embryonic development can be fascinating, especially when blastomeres are cleaving or motile cells are 'feeling' their way along. Embryonic cells seem more like active robots than passive 'bricks' [196, 249, 752, 946]. The automatic nature of development has invited comparison with programmable computers ever since the latter were invented. Indeed, two of the founders of modern computing – Alan Turing and John von Neumann – speculated extensively on embryological problems (specifically the puzzles posed by morphogenesis [907] and self-reproduction [924], respectively). Historically, the computer metaphor [14, 15, 106, 268, 423, 482, 926, 1010] has provided some valuable heuristics ... plus, unfortunately, many vacuous clichés and muddled debates [308, 605, 962]. Part of the problem is that arguments can sink into a semantic quagmire over how to define developmental 'control' [952, 954, 955] or developmental 'program' [504, 645, 826, 828] (cf. Sternberg [840] for useful definitions of both terms in the context of nematode development). Given the illustration of so many developmental mechanisms in cybernetic formats in the previous chapters, plus a new conceptual framework for thinking about patterning, there is now a fresh opportunity to reexamine the metaphor. In this chapter, an attempt is made not only to tease out new insights, but also to begin raising the metaphor to the level of a testable theory (cf. Goodwin [308] and Huszagh [405]).

Local vs. Global Information

If anatomy is indeed built by cellular 'robots', then the engineering of development might best be seen from a cell's point of view [196, 336]. Whereas a cell can 'feel' and 'taste' its neighbors and 'smell' molecules that waft in from the distance, its sense of what is happening elsewhere is necessarily limited. Its predicament is like that of an ant walking on a billboard, trying to read the message by deducing each letter from its curvature, vertex angles, and crossbars [591]. Fortunately, development seems designed for a lilliputian perspective. Thus, for example, limb regeneration (in insects and vertebrates) obeys 'Bateson's Rule' [977], which governs the arrangement and handedness of triplicated limbs that share a common shoulder or hip joint (fig. 10a). Such limbs arise spontaneously only rarely but can be easily produced by grafting operations [245] or by mutationally induced cell death [76, 283–285, 287, 749]. When two extra limbs bracket a normal limb, they have a handedness opposite to it. At the cellular level, Bateson's Rule is attributable to a 'Local Continuity Rule' [77, 510, 977] (inherent in the Polar Coordinate Model and related schemes), which requires that cells restore pattern continuity (by moving, dying, or intercalating new cells) whenever they are beside an inappropriate neighbor. Branched limbs and other regulative outcomes may seem as illogical as the equation '1 + 1 = 3', but they are perfectly natural to the participating cells because no local rules are violated, and the same holds true for other abnormal anatomies (e.g. Siamese twins) that have new internal planes of mirror symmetry [240, 311, 494, 541, 748]. Indeed, many large-scale geometries and topologies can easily be reformulated mathematically in terms of small-scale behavioral rules [2, 361, 672].

Cellular perceptions may be as confined in time as they are in space. For instance, cells that participate in multiple inductions only acquire the competence to proceed to the next step if the previous step has been completed [282]. To the extent that development proceeds by a cascade of events at the cellular level, it is 'epigenetic' [341] – i.e. genes need not be directly involved (see below). Where genes *do* seem to play crucial roles is in (1) initiating such chain reactions [307] and (2) encoding states of cellular determination [442].

Binary Codes and Boolean Logic

Digital computers use a binary code for both memory storage and information processing [934], and there is mounting evidence that cells do likewise to a limited extent. Cells can react to simple cues such as 'divide', 'move', or 'become a neuron' [196] (cf. ritualized animal behaviors that are also triggered by simple cues [25, 328, 406, 535, 562, 607]). The IF/THEN rules that link stimuli to responses are binary insofar as a cell can either

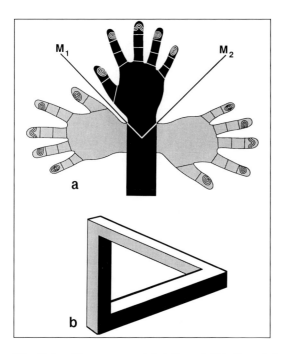

Fig. 10. Local harmony vs. global discord. This figure is intended to allow the reader to experience development from a 'cell's-eye view'. *a* Schematic diagram illustrating 'Bateson's Rule' (cf. Bateson's [46] fig. 153), as applied to the special case where three appendages (represented by hands) develop from a left-limb blastema transplanted onto a right stump (black base) in salamanders. The two extra limbs (shaded) that grow out from the graft interface are always mirror images of the central limb (black), which retains its original handedness [245]. 'M_1' and 'M_2' indicate planes of mirror symmetry. (The actual limbs that Bateson studied arose spontaneously and had both extra limbs on the *same* side of the original; cf. French [240].) Such regenerative behavior is attributable to a 'Local Continuity Rule' [77, 510, 977], which ensures that all cells ultimately (after intercalary regeneration) reside next to the types of cells that would normally be their neighbors. Notice that thumb cells reside next to other thumb cells across plane M_1, and 'little finger' cells confront one another across plane M_2. Thus, although the anatomy at a gross level is bizarre, it violates no rules from the perspective of the individual cells. After French et al. [245]. *b* 'Impossible Triangle' Illusion [680], upon which M.C. Escher based several of his famous lithographs [220, 879]. Here too, there is an obedience to rules of local connectedness, insofar as each vertex is architecturally valid, but the whole triangle is topologically heretical because its mutually orthogonal edges should not be able to achieve closure. This comparison between the triplicated limb and the Impossible Triangle is more real than metaphorical, since our retinal cells gather information in much the same *local* manner as salamander limb cells, and it is only when the higher-order centers in our visual system attempt to compile the image fragments into a self-consistent mental object that difficulties arise [501, 552]. After Escher [220].

('1') execute the behavior or ('0') not do so. The stepwise nature of differ-entiation in most organisms means that particular states constrain future choices, so that cells follow a branching pathway of binary decisions until they reach particular final states [341, 466, 796, 945, 948] (but cf. Roth [737]), at which point their 'potency' equals their 'prospective fate'. For example, the sequence of cellular decisions leading to a cholinergic motor neuron would be: 'a commitment first to ectoderm rather than to meso-derm, then to nervous tissue rather than to skin, then to neuron rather than to glia, then to motor neuron rather than to sensory neuron, and finally to synthesis of acetylcholine rather than γ-aminobutyric acid' [830].

The idea that cells might encode their determined states as ordered series of their previous choices (e.g. a final state of '11001' based upon 5 earlier decisions) was first proposed by Stuart Kauffman [438, 442, 443, 449] (cf. Slack [794]). He argued that cells could 'remember' each decision by adopting either an ON or an OFF state for individual regulatory genes (e.g. a gene being transcribed vs. not being transcribed [272, 941]), and particular combinations of the 'memory gene' products would ultimately activate other ('structural') genes that would implement the expression of the final differentiated state (e.g. hemoglobin genes in an erythrocyte).

Kauffman's original hypothesis was based upon frequencies of 'trans-determination' exhibited by cells belonging to different parts of the *Dro-sophila* body [440]. Certain types of interorgan transformation (e.g. leg cells becoming wing cells) occur more frequently than others during long-term tissue culture. If the possible transformations (leg → wing, leg → eye, eye → wing, etc.) are diagrammed as a network, then certain paths are favored over others. (Homeotic transformations in human epithelia also obey a 'Weighted Network Rule' [797].) Such biases are understandable if each binary register is controlled by a separate 'switch' gene, and mutations (or spontaneous epigenetic errors) are sufficiently rare that they typically affect only one gene at a time. For example, a code of 11111 could easily change to 10111 if the gene for the second register became defective, but two separate events would be needed to change 11111 to 10101, thereby making the first type of transformation (e.g. leg → wing) more likely than the second (e.g. leg → eye).

Combinatorial codes are efficient because they utilize a minimum number of genes to specify a maximum number of states, so evolution should have favored them [277, 443]. Genetic and molecular evidence implicates the involvement of a combinatorial code for differentiated states in *Drosophila* body segments and parasegments [7, 87, 98, 411, 414,

424, 492, 745, 853] (but cf. rebuttals [269, 298, 490]), compartments [248, 252, 443, 449] (but again cf. rebuttals [72, 331, 332]), CNS neurons [179, 181, 182], photoreceptors [713], and bristle cells [356, 369]; *C. elegans* neurons [228, 939], vulval cells [840, 845], and other cell types [363]; yeast mating types [812]; *Dictyostelium* prespore vs. prestalk cells [56]; and flowering vs. vegetative states of plant apical meristems [687]. Particular histospecific genes in various organisms are likewise thought to employ combinatorial control mechanisms [101, 175, 232, 477, 532, 880].

Convincing evidence for a combinatorial code has come from analyses of homeotic mutations that affect flower development in two distantly related plants: the crucifer *Arabidopsis thaliana* and the snapdragon *Antirrhinum majus* [67, 97, 121, 585, 698, 773]. In each of these species, there are 3 genetic functions – *a, b,* and *c* – that specify the identity of 4 types of organs (each of which occupies a separate whorl along the axis of the flower): sepals, petals, stamens, and carpels. The code has been deciphered by studying the phenotypes of double and triple mutant combinations of null alleles which cause particular organs to develop as the 'wrong' type. If the states of the *a*-, *b*-, and *c*-type genes are represented as a triplet code – with '1' indicating an active gene (or functional allele) and '0' indicating an inactive gene (or null allele) – then the codes for the various organs have been shown to be: sepals (100), petals (110), stamens (011), and carpels (001) [359].

In order for a cell (or a computer) to act upon particular combinations of 1's and 0's in a binary code, it must be capable of executing Boolean logic [305, 1010]. Continuing with the *Arabidopsis* example, if there were a regulatory gene that specifies the petal state, then it would only be expressed 'if *a* is on AND *b* is on AND *c* is off'. 'Decoding' operations of this kind are usually thought to involve multimeric complexes of *trans*-acting regulatory proteins, which control the transcription of histospecific genes by binding at upstream regulatory sites [54, 55, 162, 234, 277, 374, 676]. In the case of the 'petal gene', an upstream site(s) would presumably bind dimers of *a* and *b* proteins, and only when that site(s) is occupied would the gene be transcribed. (Evidence for regulation of *Drosophila* genes by helix-loop-helix protein dimers is discussed below.) Given that some genes have as many as 20 upstream regulatory sites, which can bind as many as 12 different trans-acting factors [880], the potential capacity of each gene to act as an information processing device becomes appreciable, leading to the notion that cells may indeed be smarter than we think [48, 162].

Default States

Surprisingly, '000' in *Arabidopsis* and *Antirrhinum* does not encode any organ of the flower, but rather specifies 'leaf': in the *abc* triple mutant, all organs of the flower are transformed into leaves [121, 585]. In his 'Metamorphosis of Plants' published in 1790, Goethe had used morphological evidence to argue that the leaf represents the 'ground state' for all floral organs [17]. A colloquial version of the Ground-State Problem is [318]: 'Is a zebra a white horse with black stripes or a black horse with white stripes?' (It is the latter [33]; cf. Carroll [98] and Lawrence [490].) Much significance has generally been attributed to such 'default' states [4, 50, 396, 845, 857], i.e. the fates that cells adopt in the absence of signals that would normally direct them into one pathway vs. another. The usual assumption is that ground-state anatomies are 'atavisms', i.e. anatomies which existed in evolutionary ancestors prior to the origin of the genes necessary to encode the 'higher' states [7, 248, 250]. Thus, when the entire bithorax complex of *Drosophila* is deleted, nearly all body segments transform into a (normally leg-bearing) second-thoracic segment [507, 787], implying that this 'centipede' anatomy is the ancestral phenotype. However, null mutations at other homeotic loci (which interact with the complex) transform nearly all segments into legless eighth-abdominal segments [169, 759, 850, 852, 855] – a wormlike condition [704]. Are both anatomies ancestral, with one more ancient than the other (cf. French [241])? Evolutionary interpretations of this sort can be misleading until more is known about both the genetic regulatory network and homologous networks in outgroup species [50].

Linguistic Hierarchies

As biologists, we are accustomed to thinking about symbolic languages and information processing [70, 303, 422, 504, 678, 1013]. Chromosomes use 'DNA language', whose alphabet contains 4 letters (the nucleotide bases), which are arranged in 3-letter words (codons), which in turn are grouped in sentences (genes) that are punctuated by start and stop codes (level I). Ribosomes then translate the text of each gene into 'protein language' (level II), whose 20 letters (amino acids) are arranged in a variety of motif domains (zinc fingers, leucine zippers, homeoboxes, etc. [10, 234, 503]) which dictate the protein's function as an enzyme, gene regulator,

structural component, etc. Control mechanisms at the level of gene regulation [439, 446, 791] (e.g. operon circuitry [305]), metabolism (e.g. activation of one enzyme via phosphorylation by another [404, 606, 875]), and
cytoarchitecture (e.g. filamin proteins cross-linking an actin network [10])
would constitute a third echelon (level III) of command structure [127],
though here the linguistic analogy weakens since there is no one-to-one
correspondence with units or sequences at the lower level [308, 645, 955,
1013].

   If there is an alphabet of development at the cellular level, then it must
consist of unitary cellular behaviors such as mitosis, movement, signaling,
adhesion, repulsion, state-change, shape-change, polarity-change, and suicide [20, 484, 663, 956]. Each of these behaviors (level IV) is implemented
by the underlying network of gene-regulatory, metabolic, or cytoarchitec-
tural factors (level III). By combining specific cellular behaviors in particular sequences, various gadgets (including any mechanism considered in
chaps. 1–6) could be built. At this echelon (level V), the 'words' would be
these modular 'subroutines' (see below), e.g. a 'word' with three 'letters'
might be: 'Divide twice, *then* rearrange in such-and-such a manner, *then*
differentiate in such-and-such a way based upon your new positions'. The
linguistic analogy regains its power at this level, since it is not merely the
combination of subunit commands but their *permutation* that matters, as
in English syntax [590]. Ultimately, the entire program of embryonic
development (level VI) would be translated into a final anatomy (level
VII). (For a different stratification scheme, cf. Ji [423].) Computer languages are designed in this same tiered fashion [878]: words in the highest-
level language (which the user employs to input instructions) are implemented by strings of operations at successively lower levels, which are
ultimately based upon a binary machine code analogous to the base
sequence in DNA. DNA may not be the only 'hard disk' inside a cell: an
intriguing recent idea is that the conformational states of tubulin dimers
within intact microtubules may function as bits for the storage of cellular
memories and the processing of cellular information [344, 345, 708].

   The hierarchical nature of development was popularized by C.H.
Waddington, whose 'epigenetic landscape' metaphor envisions developmental pathways as valleys in a landscape (fig. 11a) [380, 764, 766, 925,
926, 930]. The contours of the landscape are stabilized by an underlying
framework of interconnected guy wires, which ultimately are pegged to
individual genes. The genes function like puppeteers, with mutations
manipulating the landscape and causing cells to follow deviant paths that

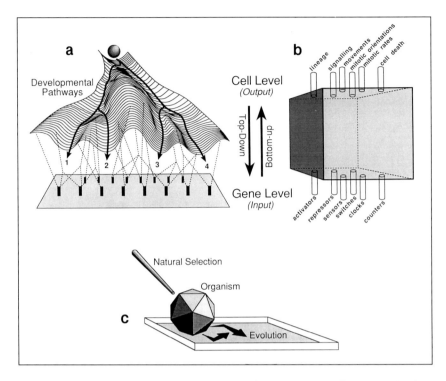

Fig. 11. Metaphors for the genetic control of development (*a, b*) and the developmental constraint of evolution (*c*). *a* Waddington's 'Epigenetic Landscape' [925, 926]. The
ball rolling downhill corresponds to a cell undergoing development. Its fate depends upon
its path. The main channel leads to a normal fate (No. 4). The contours of the canopy are
maintained by an underlying network of guy wires (= interactions of gene products)
anchored by pegs (= genes). Mutations will change the surface, resulting, in some cases, in an
abnormal fate (No. 1, 2, or 3). After Waddington [926]. *b* Conventional computer metaphor. Development is envisioned as an input/output device (or 'black box') for computing
anatomies based upon genetic information in the fertilized egg. Cellular activities (lineage,
signaling, etc.) are controlled by genetic components (activators, repressors, etc.) via an
intervening logic circuitry (not shown). Experimental approaches for analyzing developmental mechanisms have traditionally been viewed as 'top-down' (classical embryology) or
'bottom-up' (molecular genetics) [14, 68, 484, 706, 827, 830, 877, 961]. *c* The metaphor of
'Galton's Polyhedron' [317], which illustrates how the ontogeny (icosahedral cue ball) of a
species constrains its future evolution (set of possible trajectories on the pool table), given a
directional impetus from natural selection (pool cue). The 'internal tendency of an organism
to [undergo] certain *considerable* and *definite* changes' [317] in response to continuously
varying changes in its environment was the major theme of Bateson's 'Materials for the
Study of Variation' [46]. If there were no developmental constraints on evolution, then the
organism's anatomy would 'track' every environmental change in both direction and ampli-
tude (i.e. the cue ball would be spherical) [317].

lead to novel phenotypes. Paul Weiss [952, 954, 955] argued for the importance of emergent properties at all levels of the embryological hierarchy, especially during 'self-organizing' phases (cf. [308, 527]). Ergo, he reasoned, a purely reductionist molecular approach (comprising levels I–III) to development, which neglects cellular mechanisms (levels IV–VI), is bound to miss important clues that might help solve the puzzle – an opinion shared by many others [299, 349, 645, 877, 970, 997] (but cf. Caplan [96]).

Stent's Cat, Brenner's Virus, and the 'Homeobox Homunculi'

In thinking about the question of how genes 'compute' anatomy, we confront the entire span of levels from genotype to phenotype: from linear DNA to three-dimensional morphology. Whether a 'Make-a-Hand' program [69], for example, actually exists as such in our DNA depends upon which cellular mechanisms are actually involved. To the extent that the mechanisms are self-organizing and not deterministic, the correspondence between anatomical parts and genetic counterparts will be lessened [273, 309, 341, 536, 537]. In any event, the execution of a definite sequence of steps at a higher (cellular) level does not necessarily imply a colinear sequence of clustered elements (regulatory genes) at a lower level [504].

An epigenetic view of development has been advocated by two eminent developmental (*cum* molecular *cum* neuro-) biologists – Gunther Stent and Sydney Brenner – who present different clues as evidence. Stent's witness is the Siamese cat [825, 826]. The crossed eyes of this cat are attributable to a disturbance of the contralateral projection of its retinal axons, which is ultimately traceable to the same defective tyrosinase gene that gives the cat its peculiar coat color [333, 502, 781, 782]. Tyrosinase catalyzes the synthesis of the dark pigment melanin, and the retinal neurons are descended from the same cells that form the pigmented layer of the retina, which may explain their deviant afferent paths. Hence, there is apparently a cascade of errors from the mutant gene to the phenotype, but in no sense, Stent maintains, can the tyrosinase gene be imagined to 'specify' the neural circuitry of the visual system. By implication, there may be no genes whatsoever that directly specify that circuitry – or any other complicated anatomy. (However, it is unwise to try to deduce how normal genes *control* development based upon the effects of mutations that *affect* development [961, 962].) Stent cites Waddington's epigenetic landscape as the appropriate metaphor here: it is the *interactions* among *many* genes that *collectively* determine the outcome of dynamic processes at a higher (cellular) level. This same point is affirmed by Sydney Brenner, whose witness is the bacteriophage T4 [69]:

> We now know that the shape and structure of something like phage T4 is not explicitly and uniquely represented in its genome ... The key process is *self-assembly* which depends on the *bonding properties* of many different protein molecules so that the representation of the structure is *distributed* over many genes in the DNA.

There is no gene in T4, Brenner surmises, that specifies 'Make an icosahedron' [504]. Again, by extrapolation, one could imagine that the more complicated anatomies of multicellular organisms are built skyscraper-style from so many intermediate levels of interactions – each with its own emergent properties – that the edifice is virtually entirely epigenetic, soaring far above its genome.

> Development starts with a few ordered manifoldnesses; but the manifoldnesses create, by interactions, new manifoldnesses, and these are able, by acting back upon the original ones, to provoke new differences and so on. With each effect immediately a new cause is provided and the possibility of new specific action. – Driesch (1894) [185], as translated by Stern [832].
>
> Early development (may be) regarded as a series of defined morphological and physiological stages, each with its own pattern, albeit difficult to discern, and each pattern serving as the spatial condition for the transformation of that stage to the next by a limited set of morphogenic mechanisms ... There would be no long-range homing of the embryo toward the adult ... – Gerhart (1980) [274].

Opposing this argument are the 'homeobox homunculi': clusters of homeobox-containing genes in both insects and vertebrates, whose order along their respective chromosomes is colinear with the head-to-tail order of anatomical regions where they are expressed [168, 188, 260, 387, 511]. The conservation of this colinearity through some 600 million years of divergent evolution implies that the order must be important mechanistically (but cf. Holland [386]). The clusters may be positional-value 'memory boards', which record morphogen concentrations in the primary gradient field [259, 261, 403] (cf. Hanscombe et al. [348] and Sander [757]). While such one-gene-one-structure correspondences may be rare in embryos (cf. macrochaete genes [275, 743]), there is growing evidence (next section) that the genome is *functionally* (if not physically) fragmented into subunits which play multiple roles in development [1021].

Subroutines and Modules

A 'subroutine' is an algorithm that can be accessed merely by specifying its name once its sequence of instructions has been defined elsewhere [467, 934]. Subroutines can be repeatedly incorporated as modular building blocks in larger programs, which in turn can be used as units in still larger programs. An advantage of subdividing a complicated process into modular units is the efficiency it affords in the face of errors [290], as illustrated by a parable of Herbert Simon's:

> The parable concerns two watchmakers, Hora and Tempus. Both make watches consisting of a thousand parts each. Hora assembles his watches bit by bit; so when he pauses or drops a watch before it is finished, it falls to pieces and he has to start from scratch. Tempus, on the other hand, puts together sub-assemblies of ten parts each; ten of these sub-assemblies he makes into a larger sub-assembly of a hundred units; and ten of these make the whole watch. If there is a disturbance, Tempus has to repeat at worst nine assembling operations, and at best none at all. If you have a ratio of one disturbance in a hundred operations, then Hora will take four thousand times longer to assemble a watch – instead of one day, he will take eleven years [468].

Aside from error correction, the biological advantages of constructing anatomy in a modular manner [142] theoretically include: (1) economy of genetic specification, since a relatively small number of components needs to be encoded, and (2) acceleration of anatomical evolution, since changes in one subsystem need not affect other subsystems [69, 504]. A developmental subroutine would be a ritualized series of genetic or cellular actions that is used in multiple places or at multiple times during development. In the realm of morphogenesis there appear to be common subroutines for the folding of epithelia into tubes, pockets, or vesicles (regardless of which particular organ's epithelium may be involved in any given species) [30, 222, 335, 336, 628, 629, 651]. At the genetic level, the lexicon of patterning subroutines may include the following:

(1) *A neural-patterning machine.* In *Drosophila,* a single basic mechanism constructs both the embryonic central nervous system (ECNS) and the adult peripheral nervous system (APNS) [94, 369]. Two types of genes are involved, and they function sequentially. First, 'equivalence group' (EG) genes establish territories within the ectoderm (ECNS) or epidermis (APNS), presumably in response to a coordinate system of positional information [93, 276, 789]. Then, following the stochastic inception of neuroblasts (ECNS) or bristle cells (APNS) within these areas, the 'inhibitory

field' (IF) genes enable these cells to inhibit neighboring cells from differentiating in a like manner [356, 365, 369, 788]. Subsequently, the neuroblasts and bristle cells (1) delaminate from the epithelial layer, and (2) undergo a stereotyped series of mitoses leading to a clone of ganglion neurons or bristle-organ cells (shaft cell, socket cell, sheath cell, and one or more neurons), respectively [354, 355, 903]. It is noteworthy how many different schemes are concatenated in this single pathway [180, 369]: positional information, random selection, inhibitory fields, cell rearrangement, and cell lineage (though, as discussed earlier, the final cell-lineage step may not directly assign bristle-cell fates). Because the sets of EG genes used by the ECNS and APNS overlap to a large extent – as do the sets of IF genes – it is likely that the two nervous systems are built by the same machine [18, 94, 369], with the idiosyncratic features of each neuroanatomy being crafted by the unshared genes (which may 'tune' the parameters of the EG and IF subroutines to different settings).

(2) *An analog-to-digital transducer.* Some of the same genes that are used to establish equivalence groups during *Drosophila* neurogenesis (as described above) also function in embryonic segmentation and sex determination – processes that are seemingly as unrelated to each other as they are to neurogenesis. What all three systems do share is a dichotomous cellular choice: each cell must decide to become: (1) neural or non-neural; (2) part of the segment (parasegment [490, 491, 556]) or the intersegmental membrane (parasegment boundary); or (3) male or female. Although each decision is inherently 'digital' (i.e. two distinct alternative states), the factors that influence the choice may vary continuously in an analog manner. Hence, there is a need for some sort of analog-to-digital molecular transducing device. The protein products of the equivalence-group genes (*hairy, extramacrochaetae, daughterless,* and the four *achaete-scute* genes) possess a helix-loop-helix binding domain that allows them to form homo- or heterodimers [11, 100, 746] (thus integrating positive and negative analog signals), and a separate DNA-binding domain in some of them enables the dimers to bind (and regulate) other genes that are directly responsible for encoding states of cellular determination (thus converting the signals to the digital ON or OFF state of a target gene) [39, 214, 257, 258, 918]. The situation is best understood for sex determination, where each cell first measures a ratio between 'numerator' (X-chromosomal) and 'denominator' (autosomal) gene products, and then switches on or off a master gene (the *Sex-lethal* locus) based upon these inputs [218, 382, 674, 895]. (Helix-loop-helix dimerization cannot be the whole story since *runt,* a segmenta-

tion gene which lacks an HLH motif, has recently been identified as another numerator element [189]. The entire ensemble of pair-rule genes may constitute a separate analog-to-digital device [207].) The ability of a homologous vertebrate gene – MyoD – to transform fibroblasts into myoblasts [652, 942, 943] implies that the same device may be used for specifying certain cell states in vertebrates.

The above examples illustrate how sets of genes can be used repeatedly in different pathways during development. Indeed, an emerging theme in developmental biology is that small numbers of genes are often 'plugged into' large numbers of developmental circuits [845]. Cases where the same genes are successively used for different functions within a *single* pathway include: (1) the re-usage of *Drosophila* IF genes to specify cellular fates within bristle organs (e.g. a shaft cell vs. a neuron) [356, 369] and ommatidia (e.g. a cone cell vs. a pigment cell) [92]; (2) the re-usage of *Drosophila* segmentation genes to encode neural cell identities within each segment [177, 179, 181, 182, 190, 677]; and (3) the re-usage in *Arabidopsis* of one of the *c* genes (that functions in the organ identity code) to limit the number of whorls along the flower axis [121]. The efficient utilization of a limited number of genes makes sense from an evolutionary standpoint, and it helps explain why so many mutations have pleiotropic phenotypes [20, 723]. A related – and still unresolved – question is whether genomes allocate separate genes for patterning and 'housekeeping' (i.e. vital metabolic or physiological functions) [69, 70, 141, 834].

There appear to be many situations where single subroutines are employed in *similar roles at multiple locations* during development. All of them entail 'serially homologous' [136, 166, 737, 920, 932, 933] organs. As mentioned in the Introduction, such patterns are common in multicellular organisms. Their naturally occurring variants were the subject of Bateson's 'Materials for the Study of Variation' (cf. Kellogg and Bell [453] and Maynard Smith [558]). Consider your hands *versus* your feet. Because they have the same basic skeletal anatomy (as do your entire arm and leg), they may be outputs of a single subroutine, whose variables (e.g. bone lengths and articulation angles) assume different input values (see below; cf. Riedl [723] and Tabin [873]), depending upon where the limb develops [736]. Indeed, although the vertebrate limb has evolved into structures as superficially different as elephant legs and bat wings, the essential pattern of bones has remained constant [166, 312, 379, 384, 783], suggesting that the subroutine has been highly conserved while the settings of its input variables have changed drastically. If the subroutine itself were to be altered (either mutationally or evolutionarily), then the two pairs of limbs should change their morphologies coordinately. Numerous examples of such correlated changes are known [379, 419, 543, 1003]. For instance (1) the brachydactylous anomaly 'A2' in man concomitantly shortens the middle phalanges of the index finger and the second toe [1003], and (2) the panda has evolved a thumb-like sesamoid bone on both its hindfeet and forefeet, though only the bone on the forefeet has any apparent function [736]. Comparable coevolutionary trends have been found for tooth patterns in the upper *versus* the lower jaw in mammals [88]. Perhaps the clearest example of how evolution can tinker with developmental subroutines is in nematodes, where lineages vary *among* species according to the same few rules that govern differences among lineages *within* the species *C. elegans* [326, 459, 840–842].

Iteration and Halt Conditions

An intriguing aspect of many periodic patterns is their exact numbers of pattern elements [136, 233, 558]. Why, for instance, do humans typically have 10 fingers, 24 ribs, and 32 teeth (cf. the American Flag Problem)? What mechanisms ensure such constancy? This question was posed as the 'Counting Problem' by John Maynard-Smith in 1968 [559]. He envisioned two types of possible 'counting machines', which could perform an operation 'n' times: (1) 'ratio' counters, where n is the ratio between two quantities, and (2) 'digital' counters, where n is the number of interdependent events in a finite series. An example of a ratio-counting mechanism is the triggering of blastular events by the nucleocytoplasmic ratio during cleavage. Studies of haploid and polyploid embryos (which have smaller or larger nuclei, respectively) in *Xenopus, Ambystoma* (an axolotl), and *Drosophila* indicate that the number of cleavages (and the onset of transcription known as the 'mid-blastula transition') are determined by the nucleocytoplasmic ratio [206, 319, 399, 460, 464, 639, 640, 762]. The most popularly hypothesized *digital* counting devices are 'chemical counters' [558], where different gene products are synthesized at each step of a process (A, B, C, etc.) until the last product terminates the process. Such models have been invoked to explain the number of cleavages in *Volvox* [463, 869, 870]; the numbers of segments in horseshoe crabs [416], short-germ insects [757], and leeches [210] (including leech segmental identity [553]); and the timing of embryonic events in general (with or without a causal link to mitosis) [760, 761] (cf. Cooke and Smith [145]). (As for how mitotic counters might trigger differentiation events, cf. Temple and Raff [876] and the flowchart in figure 16–38 of Alberts et al. [10].)

What would happen if the counting mechanism for an iterative process were to malfunction because of a failure of its 'halt condition' (i.e. the rule that dictates 'IF condition x prevails THEN stop!')? As programmers alas know well, such default 'bugs' can cause repetitions to occur ad infinitum. Mutations that cause comparable 'infinite loop' phenotypes include: *unc-86* mutations in *C. elegans,* which force certain lineages to repeat their pedigree rules many times over [104, 228, 229] (cf. related nematode mutations [12, 750]); the *floricaula* mutation in the snapdragon *Antirrhinum* [120, 122], which replaces flowers with iterated inflorescence meristems (cf. comparable mutations in other plants [871]); and *bag-of-marbles* mutations in *Drosophila* [565], which increase the number of cells per ovarian cyst from 16 (a precise number in wild-type flies, which is due to exactly 4 mitoses per germ cell) to between 50 and several hundred (cf. Lifschytz [517]). Such 'bugs' may have been instrumental in the evolutionary origin of extremely high numbers of vertebrae in snakes and body segments in millipedes.

'Counters' are only applicable to patterns whose elements are specified sequentially. The issue of how element number is controlled in 'synchronic' [927] patterns would seem to call for entirely different mechanisms [368, 558, 728] (but cf. Meinhardt [573]). As discussed in chapter 1, positional information models are designed to ensure constant (size-independent) patterns, including fixed numbers of pattern elements. For prepattern mechanisms, some authors have envisioned the triggering of the prepattern algorithm only when the developing organ reaches a critical size [134, 136, 138, 1008]. For determination wave mechanisms, the rate of the wave movement along the tissue could be established earlier by a positional information scheme [146]. Interestingly, there are patterns whose element number varies along one axis but not another. Examples include mouse whiskers [191] and fly bristles [367]. In the latter case, the number of rows of bristles is size-invariant, whereas the number of bristles within each row varies in proportion to the size of the organ [367].

Input-Output, 'Morphospaces', and Evolutionary Constraints

The notion that developmental pathways can constrain anatomical evolution was championed by D'Arcy Thompson [882] (e.g. his 'grid transformations') and Richard Goldschmidt [295] (e.g. his 'norm of reactivity'; cf.

Kauffman [444]). It has experienced a revival lately, sparked by Stephen Gould's book 'Ontogeny and Phylogeny' [315] and fueled by Alberch et al.'s [9] 'ontogenetic trajectory' technique for operational analysis [466, 608, 704]. A vivid illustration of the relationship between development and evolution is provided by Raup's studies of the shapes of mollusc shells [709, 710, 712]. Only three variables (the rate of growth of the shell's mouth, the mouth's distance from an imaginary axis around which it revolves, and the mouth's rate of translation along that axis) are required to simulate most shell shapes. The variables define a 3-dimensional 'morphospace' [316, 560] within which the actual shapes of living and fossil families of molluscs can be plotted as contiguous domains [711]. Evidently, all molluscs share an ancient 'SHELL' algorithm, which can produce a variety of output shapes (e.g. corkscrew, planar spiral, or bivalve), given different species-specific settings of the input variables. Shapes outside the space should never arise unless the mechanism itself is altered. Moreover, if each variable is under separate genetic control, then the most likely pathways of evolutionary change from any given point in the space would be along vectors parallel to the axes.

Comparable morphospaces for *patterns* can be imagined for many of the mechanisms discussed earlier [646, 718, 840]. One example would be the morphospace of possible phyllotactic patterns (spiral, distichous, and whorled) defined by inhibitory field [593, 721, 883] or physical force [325] models. A generic evolutionary trend that is expected for any mechanism which lacks the ability to regulate (e.g. prepattern devices) is a proportionality between the number or pattern elements and the size of the pattern [367] (chap. 3).

Positional information mechanisms are not usually helpful in predicting how patterns should change with evolution [540, 1002]. However, one prediction concerning digit patterns has led to some provocative questions. A complication in trying to apply the Polar Coordinate Model [81, 245] (chap. 1) to vertebrate limbs (vs. insect appendages) is that the coordinate system is supposedly centered on the tip of the limb, but the tip branches into digits, which raises two questions: (1) which digit, if any, lies at the origin, and (2) what role, if any, does the coordinate system play in specifying digit positions? Based upon the results of digit amputation experiments with newt limbs, Stock and Bryant [848] proposed that (1) none of the digits resides at the origin, and (2) each prospective digit inherits a different subset of coordinates which are lateral to the origin. Because the subsets contain less than half of the circumferential coordinates, they automatically (due to the Shortest Intercalation Rule) undergo duplication. Such duplication explains why vertebrate fingers are internally mirror-symmetric (i.e. the left half of each finger is a mirror image of the right half). Moreover, because those digits closest to the origin would contain more coordinates, they should grow to greater length. Hence, the 'Stock-Bryant Rule' for digit evolution dictates that the lengths of the digits within a limb must decrease linearly from a single 'high point' (defined by the origin of the coordinate system). That point could be centrally located (e.g. the human hand, where the middle finger is longest) or more lateral (e.g. the human foot, where the longest digit is typically the hallux), but there should never be a pattern where two long digits flank a shorter one (since this would imply two high points) [848]. Contrary to expectation, such patterns do exist, though they are rare [383, 627]. Alternative models, based upon sequentially branching condensations of cartilage cells, have been proposed by Oster et al. [660, 663, 783] and others [636, 638, 966], and they predict an entirely different set of developmental constraints [8, 208, 619, 663, 783]. The reader should consult Gardiner and Bryant [253] for what may be a definitive answer to the classical problem of the 'metapterygial axis' and the puzzle of 'Gregory's Pyramid'.

Game Theory

Every scientist dreams of discovering a universal law of nature that could crystallize a collection of seemingly unconnected facts into a glorious gem of insight. Biologists have been frustrated in this quest because living beings, unlike atoms, are only comprehensible in the context of their histories, and different species have evolved differently [272, 417, 563, 946]. There is, unfortunately, no periodic chart of cell types, nor any equivalent of '$F = ma$' that governs developmental trajectories. However – and here is where hope may still lie for some fulfillment – *there may be a finite number of elemental patterning strategies* [162, 489]. The rainbow spectrum of species-specific developmental pathways could then be unwoven into a limited number of primary strands. The collection of schemes surveyed in the earlier chapters, together with the kinds of devices discussed in this chapter, may constitute a large portion of the contraptions that evolution has invented to solve the problem of embryo engineering. If that problem is envisioned as a game, then this book represents an attempt to devise a pluralistic 'game theory'. Even if all the models and metaphors cannot untie development's Gordian Knot, they at least portray *possible* solutions that can inspire us as we grope.

References[1]

1   Abbott LA, Sprey TE: Components of positional information in the developing wing margin of the *Lyra* mutant of *Drosophila*. Roux's Arch Dev Biol 1990;198:448–459.

2   Abelson H, diSessa AA: Turtle Geometry: The Computer as a Medium for Exploring Mathematics. Cambridge, MIT Press, 1980.

3   Ada GL, Nossal G: The clonal-selection theory. Sci Am 1987;257:62–69.

4   Adler R, Hatlee M: Plasticity and differentiation of embryonic retinal cells after terminal mitosis. Science 1989;243:391–393.

5   Akam M: The molecular basis for metameric pattern in the *Drosophila* embryo. Development 1987;101:1–22.

6   Akam M: Making stripes inelegantly. Nature 1989;341:282–283.

7   Akam M, Dawson I, Tear G: Homeotic genes and the control of segment diversity. Development 1988;104(suppl):123–133.

8   Alberch P: Developmental constraints: Why St. Bernards often have an extra digit and poodles never do. Am Nat 1985;126:430–433.

9   Alberch P, Gould SJ, Oster GF, Wake DB: Size and shape in ontogeny and phylogeny. Paleobiology 1979;5:296–317.

10 Alberts B, Bray D, Lewis J, Raff M, Roberts K, Watson JD: Molecular Biology of the Cell, ed. 2. New York, Garland, 1989.

11 Alonso MC, Cabrera CV: The *achaete-scute* gene complex of *Drosophila melanogaster* comprises four homologous genes. EMBO J 1988;7:2585–2591.

12 Ambros V, Horvitz HR: The *lin-14* locus of *Caenorhabditis elegans* controls the time of expression of specific postembryonic developmental events. Genes Dev 1987;1:398–414.

13 Anderson DT: Embryology and Phylogeny in Annelids and Arthropods. New York, Pergamon Press, 1973.

14 Apter MJ: Cybernetics and Development. London, Pergamon Press, 1966.

15 Apter MJ, Wolpert L: Cybernetics and development. I. Information theory. J Theor Biol 1965;8:244–257.

16 Araki S: Dynamics of planetary rings. Am Sci 1991;79:44–59.

17 Arber A: Goethe's botany. Chron Bot 1946;10:63–126.

18 Artavanis-Tsakonas S, Simpson P: Choosing a cell fate: A view from the *Notch* locus. Trends Genet 1991;7:403–408.

19 Ashburner M: Puffs, genes, and hormones revisited. Cell 1990;61:1–3.

[1] *Note added in proof:* Excellent reviews of patterning mechanisms have recently been published in Cell 1992;68:185–322.

20   Atchley WR, Hall BK: A model for development and evolution of complex morpho-
     logical structures. Biol Rev 1991;66:101–157.
21   Atkeson CG: Learning arm kinematics and dynamics. Ann Rev Neursosci 1989;12:
     157–183.
22   Babloyantz A: Self-organization phenomena resulting from cell-cell contact. J Theor
     Biol 1977;68:551–561.
23   Babloyantz A, Hiernaux J: Models for positional information and positional differ-
     entiation. Proc Natl Acad Sci USA 1974;71:1530–1533.
24   Babloyantz A, Hiernaux J: Models for cell differentiation and generation of polarity
     in diffusion-governed morphogenetic fields. Bull Math Biol 1975;37:637–657.
25   Baerends GP: The functional organization of behaviour. Anim Behav 1976;24:726–
     738.
26   Bak P, Chen K, Creutz M: Self-organized criticality in the 'Game of Life'. Nature
     1989;342:780–782.
27   Baker NE, Mlodzik M, Rubin GM: Spacing differentiation in the developing *Dro-
     sophila* eye: A fibrinogen-related lateral inhibitor encoded by *scabrous.* Science
     1990;250:1370–1377.
28   Banerjee U, Zipursky SL: The role of cell-cell interaction in the development of the
     Drosophila visual system. Neuron 1990;4:177–187.
29   Bang AG, Hartenstein V, Posakony JW: *Hairless* is required for the development of
     adult sensory organ precursor cells in *Drosophila.* Development 1991;111:89–104.
30   Bard J: Morphogenesis: The Cellular and Molecular Processes of Developmental
     Anatomy. Cambridge, Cambridge University Press, 1990.
31   Bard J, Lauder I: How well does Turing's theory of morphogenesis work? J Theor
     Biol 1974;45:501–531.
32   Bard JBL: A unity underlying the different zebra striping patterns. J Zool [Lond]
     1977;183:527–539.
33   Bard JBL: A model for generating aspects of zebra and other mammalian coat pat-
     terns. J Theor Biol 1981;93:363–385.
34   Bard JBL: Butterfly wing patterns: How good a determining mechanism is the sim-
     ple diffusion of a single morphogen? J Embryol Exp Morphol 1984;84:255–274.
35   Bard JBL: A traction-based mechanism for somitogenesis in the chick. Roux's Arch
     Dev Biol 1988;197:513–517.
36   Bard JBL: Traction and the formation of mesenchymal condensations in vivo. Bio-
     Essays 1990;12:389–395.
37   Bard JBL, Ross ASA: The morphogenesis of the ciliary body of the avian eye. I.
     Lateral cell detachment facilitates epithelial folding. Dev Biol 1982;92:73–86.
38   Bard JBL, Ross ASA: The morphogenesis of the ciliary body of the avian eye. II.
     Differential enlargement causes an epithelium to form radial folds. Dev Biol 1982;
     92:87–96.
39   Barinaga M: Dimers direct development. Science 1991;251:1176–1177.
40   Barnsley MF, Devaney RL, Mandelbrot BB, Peitgen H-O, Saupe D, Voss RF: The
     Science of Fractal Images. New York, Springer, 1988.
41   Barthalay Y, Hipeau-Jacquotte R, de la Escalera S, Jiménez F, Piovant M: *Drosoph-
     ila* neurotactin mediates heterophilic cell adhesion. EMBO J 1990;9:3603–3609.
42   Basler K, Hafen E: Specification of cell fate in the developing eye of *Drosophila.*
     BioEssays 1991;13:621–631.

43   Bastiani MJ, Doe CQ, Helfand SL, Goodman CS: Neuronal specificity and growth
      cone guidance in grasshopper and *Drosophila* embryos. Trends Genet 1985;8:257–
      266.

44   Bate M, Martinez Arias A: The embryonic origin of imaginal discs in *Drosophila.*
      Development 1991;112:755–761.

45   Bates CA, Killackey HP: The organization of the neonatal rat's brainstem trigeminal
      complex and its role in the formation of central trigeminal patterns. J Comp Neurol
      1985;240:265–287.

46   Bateson W: Materials for the Study of Variation. London, MacMillan, 1894.

47   Baumgartner S, Noll M: Network of interactions among pair-rule genes regulating
      paired expression during primordial segmentation of *Drosophila.* Mech Dev 1991;
      33:1–18.

48   Beardsley T: Smart genes. Sci Am 1991;265:86–95.

49   Beeman RW: A homoeotic gene cluster in the red flour beetle. Nature 1987;327:
      247–249.

50   Beeman RW, Stuart JJ, Haas MS, Denell RE: Genetic analysis of the homeotic gene
      complex (HOM-C) in the beetle *Tribolium castaneum.* Dev Biol 1989;133:196–
      209.

51   Belic MR, Skarka V, Deneubourg JL, Lax M: Mathematical model of honeycomb
      construction. J Math Biol 1986;24:437–449.

52   Bellairs R: The mechanism of somite segmentation in the chick embryo. J Embryol
      Exp Morph 1979;51:227–243.

53   Bellairs R, Curtis ASG, Sanders EJ: Cell adhesiveness and embryonic differentia-
      tion. J Embryol Exp Morphol 1978;46:207–213.

54   Benfey PN, Chua N-H: The cauliflower mosaic virus 35S promoter: combinatorial
      regulation of transcription in plants. Science 1990;250:959–966.

55   Berk AJ, Schmidt MC: How *do* transcription factors work? Genes Dev 1990;4:151–
      155.

56   Berks M, Kay RR: Combinatorial control of cell differentiation by cAMP and DIF-1
      during development of *Dictyostelium discoideum.* Development 1990;110:977–
      984.

57   Berlot J, Goodman CS: Guidance of peripheral pioneer neurons in the grasshopper:
      adhesive hierarchy of epithelial and neuronal surfaces. Science 1984;223:493–495.

58   Berridge MJ, Rapp PE, Treherne JE (eds): Cellular oscillators. J Exp Biol 1979;
      81.

59   Bissen ST: Cell interactions in the developing leech embryo. BioEssays 1986;4:152–
      157.

60   Bissen ST: Early differences between alternate n blast cells in leech embryo. J Neu-
      robiol 1987;18:251–269.

61   Bizzi E, Mussa-Ivaldi FA, Giszter S: Computations underlying the execution of
      movement: a biological perspective. Science 1991;253:287–291.

62   Bloch R: Histological foundations of differentiation and development in plants; in
      Ruhland W (ed): Handbuch der Pflanzenphysiologie. Berlin, Springer, 1965,
      pp 146–188.

63   Blochlinger K, Jan LY, Jan YN: Transformation of sensory organ identity by ectopic
      expression of Cut in *Drosophila.* Genes Dev 1991;5:1124–1135.

64   Bode PM, Bode HR: Formation of pattern in regenerating tissue pieces of *Hydra*

attenuata. IV. Three processes combine to determine the number of tentacles. Development 1987;1:89–98.

65 Bodmer R, Barbel S, Sheperd S, Jack JW, Jan LY, Jan YN: Transformation of sensory organs by mutations of the cut locus of *D. melanogaster.* Cell 1987;51: 293–307.

66 Booker R, Truman JW: *Octopod,* a homeotic mutation of the moth *Manduca sexta,* influences the fate of identifiable pattern elements within the CNS. Development 1989;105:621–628.

67 Bowman JL, Smyth DR, Meyerowitz EM: Genetic interactions among floral homeotic genes of *Arabidopsis.* Development 1991;112:1–20.

68 Brenner S: Closing remarks: The genetic outlook; in Porter R, Rivers J (eds): Cell Patterning. Ciba Found. Symp. (N.S.), vol 29. New York, Elsevier, 1975, pp 343–345.

69 Brenner S: Genes and development; in Lloyd CW, Rees DA (eds): Cellular Controls in Differentiation. New York, Academic Press, 1981, pp 3–7.

70 Brenner S, Dove W, Herskowitz I, Thomas R: Genes and development: Molecular and logical themes. Genetics 1990;126: 479–486.

71 Britton NF: Reaction-Diffusion Equations and Their Applications to Biology. New York, Academic Press, 1986.

72 Brower DL: The sequential compartmentalization of *Drosophila* segments revisited. Cell 1985;41:361–364.

73 Brower DL: *engrailed* gene expression in *Drosophila* imaginal discs. EMBO J 1986;5: 2649–2656.

74 Brown TH, Kairiss EW, Keenan CL: Hebbian synapses: Biophysical mechanisms and algorithms. Ann Rev Neurosci 1990;13:475–511.

75 Bryant PJ: Localized cell death caused by mutations in a *Drosophila* gene coding for a transforming growth factor-β homolog. Dev Biol 1988;128:386–395.

76 Bryant PJ, Girton JR: Genetics of pattern formation; in Siddiqi O, Babu P, Hall LM, Hall JC (eds): Development and Neurobiology of *Drosophila.* New York, Plenum Press, 1980, pp 109–127.

77 Bryant PJ, Girton JR, Martin P: Physical and pattern continuity in the insect epidermis; in Locke M, Smith DS (eds): Insect Biology in the Future. New York, Academic Press, 1980, pp 517–542.

78 Bryant PJ, Huettner B, Held LI, Jr, Ryerse J, Szidonya J: Mutations at the *fat* locus interfere with cell proliferation control and epithelial morphogenesis in *Drosophila.* Dev Biol 1988;129:541–554.

79 Bryant PJ, Levinson P: Intrinsic growth control in the imaginal primordia of *Drosophila,* and the autonomous action of a lethal mutation causing overgrowth. Dev Biol 1985;107:355–363.

80 Bryant PJ, Simpson P: Intrinsic and extrinsic control of growth in developing organs. Q Rev Biol 1984;59:387–415.

81 Bryant SV, Bryant PJ, French V: Distal regeneration and symmetry. Science 1981; 212:993–1002.

82 Budrene EO, Berg HC: Complex patterns formed by motile cells of *Escherichia coli.* Nature 1991;349:630–633.

83 Buikema WJ, Haselkorn R: Characterization of a gene controlling heterocyst differentiation in the cyanobacterium *Anabaena* 7120. Genes Dev 1991;5:321–330.

84 Bünning E: Die Entstehung von Mustern in der Entwicklung von Pflanzen; in Ruhland W (ed): Handbuch der Pflanzenphysiologie, vol 15. Berlin, Springer, 1965, pp 383–408.

85 Bünning E, Sagromsky H: Die Bildung des Spaltöffnungsmusters in der Blattepidermis. Z Naturforsch 1948;3b:203–216.

86 Burgess EA, Duncan I: Direct control of antennal identity by the *spineless-aristapedia* gene of *Drosophila*. Mol Gen Genet 1990;221:347–352.

87 Busturia A, Casanova J, Sánchez-Herrero E, Morata G: Structure and function of the bithorax complex genes of *Drosophila;* in Evered D, Marsh J (eds): Cellular Basis of Morphognesis. Ciba Found Symp, vol 144. New York, Wiley, 1989, pp 227–242.

88 Butler PM: Studies of the mammalian dentition. I. The teeth of *Centetes ecaudatus* and its allies. Proc Zool Soc Lond [B] 1937:107:103–132.

89 Butler PM: The ontogeny of mammalian heterodonty. J Biol Bucc 1978;6:217–227.

90 Byrne G, Cox EC: Spatial patterning in *Polysphondylium:* Monoclonal antibodies specific for whorl prepatterns. Dev Biol 1986;117:442–455.

91 Cagan RL, Ready DF: The emergence of order in the *Drosophila* pupal retina. Dev Biol 1989;136:346–362.

92 Cagan RL, Ready DF: *Notch* is required for successive cell decisions in the developing *Drosophila* retina. Genes Dev 1989;3:1099–1112.

93 Campos-Ortega JA: Mechanisms of a cellular decision during embryonic development of *Drosophila melanogaster:* Epidermogenesis or neurogenesis. Adv Genet 1990;27:403–453.

94 Campos-Ortega JA, Jan YN: Genetic and molecular bases of neurogenesis in *Drosophila melanogaster.* Ann Rev Neurosci 1991;14:399–420.

95 Caplan AI, Fiszman MY, Eppenberger HM: Molecular and cell isoforms during development. Science 1983;221:921–927.

96 Caplan AL: Rehabilitating reductionism. Am Zool 1988;28:193–203.

97 Carpenter R, Coen ES: Floral homeotic mutations produced by transposon-mutagenesis in *Antirrhinum majus.* Genes Dev 1990;4:1483–1493.

98 Carroll SB: Zebra patterns in fly embryos: Activation of stripes or repression of interstripes? Cell 1990;60:9–16.

99 Carroll SB, Vavra SH: The zygotic control of *Drosophila* pair-rule gene expression. II. Spatial repression by gap and pair-rule gene products. Development 1989;107: 673–683.

100 Caudy M, Vässin H, Brand M, Tuma R, Jan LY, Jan YN: *daughterless,* a *Drosophila* gene essential for both neurogenesis and sex determination, has sequence similarities to *myc* and the *achaete-scute* complex. Cell 1988;55:1061–1067.

101 Cavener DR: Combinatorial control of structural genes in *Drosophila:* Solutions that work for the animal. BioEssays 1987;7:103–107.

102 Caveney S: Cell communication and pattern formation in insects; in Locke M, Smith ES (eds): Insect Biology in the Future. New York, Academic Press, 1980, pp 565–582.

103 Celis JFD, Marí-Beffa M, García-Bellido A: Function of trans-acting genes of the *achaete-scute* complex in sensory organ patterning in the mesonotum of *Drosophila.* Roux's Arch Dev Biol 1991;200:64–76.

104 Chalfie M, Horvitzh HR, Sulston JE: Mutations that lead to reiterations in the cell lineages of *C. elegans.* Cell 1981;24:59–69.

105 Chandebois R: Cell sociology and the problem of position effect: pattern formation, origin and role of gradients. Acta Biotheor 1977;26:203–238.

106 Chandebois R, Faber J: Automation in Animal Development. Monogr Dev Biol, vol 16. Basel, Karger, 1983.

107 Charles AC, Merrill JE, Dirksen ER, Sanderson MJ: Intercellular signaling in glial cells: Calcium waves and oscillations in response to mechanical stimulation and glutamate. Neuron 1991;6:983–992.

108 Cheney CM, Lash JW: An increase in cell-cell adhesion in the chick segmental plate results in a meristic pattern. J Embryol Exp Morphol 1984;79:1–10.

109 Child CM: Physiological dominance and physiological isolation in development and reconstitution. Roux Arch Entw-Mech Org 1929;117:21–66.

110 Child CM: Patterns and Problems of Development. Chicago, University of Chicago Press, 1941.

111 Chomsky N: Language and Mind. New York, Harcourt Brace Jovanovich, 1968.

112 Clarke PGH: Chance, repetition, and error in the development of normal nervous systems. Perspect Biol Med 1981;25:2–19.

113 Clarke PGH: Developmental cell death: Morphological diversity and multiple mechanisms. Anat Embryol 1990;181:195–213.

114 Claxton JH: The determination of patterns with special reference to that of the central primary skin follicles in sheep. J Theor Biol 1964;7:302–317.

115 Claxton JH: Some quantitative features of *Drosophila* sterinte bristle patterns. Aust J Biol Sci 1974;27:533–543.

116 Claxton JH: Developmental origin of even spacing between the microchaetes of *Drosophila melanogaster.* Aust J Biol Sci 1976;29:131–135.

117 Claxton JH, Sholl CA: A model of pattern formation in the primary skin follicle population of sheep. J Theor Biol 1973;40:353–367.

118 Cockayne EA: Homoeosis and heteromorphosis in insects. Trans Ent Soc Lond 1926;74:203–230.

119 Codd EF: Cellular Automata. New York, Academic Press, 1968.

120 Coen ES, Doyler S, Romero JM, Elliott R, Margrath R, Carpenter R: Homeotic genes controlling flower development in *Antirrhinum.* Development 1991;(suppl)1:149–155.

121 Coen ES, Meyerowitz EM: The war of the whorls: genetic interactions controlling flower development. Nature 1991;353:31–37.

122 Coen ES, Romero JM, Doyle S, Elliott R, Murphy G, Carpenter R: *floricaula:* A homeotic gene required for flower development in Antirrhinum majus. Cell 1990;63:1311–1322.

123 Cohen B, Wimmer EA, Cohen SM: Early development of leg and wing primordia in the *Drosophila* embryo. Mech Dev 1991;33:229–240.

124 Cohen MH: Models for the control of development. Symp Soc Exp Biol 1971;25:455–476.

125 Cohen MH: Models of clocks and maps in developing organisms; in Lectures on Mathematics in the Life Sciences, vol 3. Providence, American Mathematics Society, 1972, pp 1–32.

126 Cohen SM: Specification of limb development in the *Drosophila* embryo by posi-
     tional cues from segmentation genes. Nature 1990;343:173–177.

127 Collado-Vides J: Towards a grammatical paradigm for the study of the regulation of
     gene expression; in Goodwin B, Saunders P (eds): Theoretical Biology: Epigenetic
     and Evolutionary Order from Complex Systems. Edinburgh, Edinburg University
     Press, 1989, pp 211–224.

128 Constantine-Paton M: The retinotectal hookup: The process of neural mapping; in
     Subtelny S, Green PB (eds): Developmental Order: Its Origin and Regulation. Symp
     Soc Dev Biol, vol 40. New York, Liss, 1982, pp 317–349.

129 Constantine-Paton M, Blum AS, Mendez-Otero R, Barnstable CJ: A cell surface
     molecule distributed in a dorsoventral gradient in the perinatal rat retina. Nature
     1986;324:459–462; cf. erratum in Nature 325:284.

130 Constantine-Paton M, Cline HT, Debski EA: Neural activity, synaptic convergence,
     and synapse stabilization in the developing central nervous system; in Landmeser
     LT (ed): The Assembly of the Nervous System. Symp Soc Dev Biol, vol 47. New
     York, Liss, 1989, pp 279–300.

131 Constantine-Paton M, Law MI: The development of maps and stripes in the brain.
     Sci Am 1982;247:62–70.

132 Cook TA: The Curves of Life. London, Constable, 1914.

133 Cooke J: Control of somite number during morphogenesis of a vertebrate, *Xenopus
     laevis.* Nature 1975;254:196–199.

134 Cooke J: The emergence and regulation of spatial organization in early animal
     development. Ann Rev Biophys Bioeng 1975;4:185–217.

135 Cooke J: Experimental analysis and a theory of the control of somite number during
     amphibian morphogenesis; in McMahon D, Fox CF (eds): Developmental Biology:
     Pattern Formation, Gene Regulation. ICN-UCLA Symp Molec Cell Biol, vol 2.
     Menlo Park, Benjamin, 1975, pp 205–226.

136 Cooke J: The problem of periodic patterns in embryos. Phil Trans R Soc Lond [B]
     1981;295:509–524.

137 Cooke J: Scale of body pattern adjusts to available cell number in amphibian
     embryos. Nature 1981;290:775–778.

138 Cooke J: The relation between scale and the completeness of pattern in vertebrate
     embryogenesis: Models and experiments. Am Zool 1982;22:91–104.

139 Cooke J: Evidence for specific feedback signals underlying pattern control during
     vertebrate embryogenesis. J Embryol Exp Morphol 1983;76:95–114.

140 Cooke J: Morphallaxis and early vertebrate development; in Malacinski GM, Bryant
     SV (eds): Pattern Formation: A Primer in Developmental Biology. New York, Mac-
     millan, 1984, pp 481–506.

141 Cooke J: Vertebrate development through a glass darkly. BioEssays 1986;4:185–
     186.

142 Cooke J: The early embryo and the formation of body pattern. Am Sci 1988;76:
     35–41.

143 Cooke J, Elsdale T: Somitogenesis in amphibian embryos. III. Effects of ambient
     temperature and of developmental stage upon pattern abnormalities that follow
     short temperature shocks. J Embryol Exp Morphol 1980;58:107–118.

144 Cooke J, Goodwin BC: Periodic wave propagation and pattern formation: applica-

tion to problems in development; in Lectures on Mathematics in the Life Sciences, vol 3, Providence, American Mathematics Society, 1972, pp 33–60.

145 Cooke J, Smith JC: Measurement of developmental time by cells of early embryos. Cell 1990;60:891–894.

146 Cooke J, Zeeman EC: A clock and wavefront model for control of the number of repeated structures during animal morphogenesis. J Theor Biol 1976;58:455–476.

147 Cornell-Bell AH, Finkbeiner SM, Cooper MS, Smith SJ: Glutamate induces calcium waves in cultured astrocytes: Long range glial signaling. Science 1990;247:470–473.

148 Cowan WM, Fawcett JW, O'Leary DDM, Stanfield BB: Regressive events in neurogenesis. Science 1984;225:1258–1265.

149 Crick F: Diffusion in embryogenesis. Nature 1970;225:420-422.

150 Crick FHC: Thinking about the brain. Sci Am 1979;241:219–232.

151 Crick FHC, Lawrence PA: Compartments and polyclones in insect development. Science 1975;189:340–347.

152 Croxdale J, Smith J, Yandell B, Johnson JB: Stomatal patterning in *Tradescantia:* an evaluation of the cell lineage theory. Dev Biol 1992;149:158–167.

153 Cummings FW, Prothero JW: A model of pattern formation in multicellular organisms. Collective Phenom 1978;3:41–53.

154 Cummins H, Midlo C: Finger Prints, Palms and Soles. An Introduction to Dermatoglyphics. New York, Dover, 1943.

155 Cutter EG, Feldman LJ: Trichoblasts in *Hydrocharis.* I. Origin, differentiation, dimensions and growth. Am J Bot 1970;57:190–201.

156 Czihak G: Echinoids; in Reverberi G (ed): Experimental Embryology of Marine and Freshwater Invertebrates. Amsterdam Elsevier, 1971, pp 363–506.

157 Dalcq AM: Form and Causality in Early Development. Cambridge, Cambridge University Press, 1938.

158 Davidson D: The mechanism of feather pattern development in the chick. I. The time of determination of feather position. J Embryol Exp Morphol 1983;74:245–259.

159 Davidson D: The mechanism of feather pattern development in the chick. II. Control of the sequence of pattern formation. J Embryol Exp Morphol 1983;74:261–273.

160 Davidson EH: Gene Activity in Early Development, ed 3. New York, Academic Press, 1986.

161 Davidson EH: Lineage-specific gene expression and the regulative capacities of the sea urchin embryo: A proposed mechanism. Development 1989;105:421–445.

162 Davidson EH: How embryos work: A comparative view of diverse models of cell fate specification. Development 1990;108:365–389.

163 Davidson EH: Spatial mechanisms of gene regulation in metazoan embryos. Development 1991;113:1–26.

164 Davies JA, Cook GMW: Growth cone inhibition – an important mechanism in neural development? BioEssays 1991;13:11–15.

165 Davis I, Ish-Horowicz D: Apical localization of pair-rule transcripts requires 3′ sequences and limits protein diffusion in the *Drosophila* blastoderm embryo. Cell 1991;67:927–940.

166 de Beer GR: Homology, an unsolved problem. Oxford Biol Readers 1971;11:1–16.

167 De Reffye P, Lecoustre R, Edelin C, Dinouard P: Modelling plant growth and architecture; in Goldbeter A (ed): Cell to Cell Signalling: From Experiments to Theoretical Models. New York, Academic Press, 1989, pp 237–246.

168 De Robertis EM, Oliver G, Wright CVE: Homeobox genes and the vertebrate body plan. Sci Am 1990;263:46–52.

169 Denell RE, Frederick RD: Homoeosis in *Drosophila*: A description of the Polycomb lethal syndrome. Dev Biol 1983;97:34–47.

170 Deuchar EM, Burgess AMC: Somite segmentation in amphibian embryos: is there a transmitted control mechanism? J Embryol Exp Morphol 1967;17:349–358.

171 Devreotes PN: Chemotaxis; in Loomis WF (ed): The Development of *Dictyostelium discoideum*. New York, Academic Press, 1982, pp 117–168.

172 Devreotes PN: Chemotaxis in eukaryotic cells: A focus on leukocytes and *Dictyostelium*. Ann Rev Cell Biol 1988;4:649–686.

173 Dewdney AK: The hodgepodge machine makes waves. Sci Am 1988;259:104–107.

174 Dewdney AK: The Turing Omnibus: 61 Excursions in Computer Science. Rockville, Computer Science Press, 1989.

175 Dickinson WJ: On the architecture of regulatory systems: Evolutionary insights and implications. BioEssays 1988;8:204–208.

176 Dodd J, Jessell TM: Axon guidance and the patterning of neuronal projections in vertebrates. Science 1988;242:692–699.

177 Doe CQ, Chu-LaGraff Q, Wright DM, Scott MP: The *prospero* gene specifies cell fates in the *Drosophila* central nervous system. Cell 1991;65:451–464.

178 Doe CQ, Goodman CS: Early events in insect neurogenesis. II. The role of cell interactions and cell lineage in the determination of neuronal precursor cells. Dev Biol 1985;111:206–219.

179 Doe CQ, Hiromi Y, Gehring WJ, Goodman CS: Expression and function of the segmentation gene *fushi tarazu* during *Drosophila* neurogenesis. Science 1988;239:170–175.

180 Doe CQ, Kuwada JY, Goodman CS: From epithelium to neuroblasts to neurons: the role of cell interactions and cell lineage during insect neurogenesis. Phil Trans R Soc Lond [B] 1985;312:67–81.

181 Doe CQ, Smouse D, Goodman CS: Control of neuronal fate by *Drosophila* segmentation gene *even-skipped*. Nature 1988;333:376–378.

182 Doe CQ, Smouse DT: The origins of cell diversity in the insect central nervous system. Semin Cell Biol 1990;1:211–218.

183 Dohle W, Scholtz G: Clonal analysis of the crustacean segment: The discordance between genealogical and segmental borders. Development 1988;104(suppl):147–160.

184 Dosier LW, Riopel JL: Origin, development, and growth of differentiating trichoblasts in *Elodea canadensis*. Am J Bot 1978;65:813–822.

185 Driesch H: Analytische Theorie der organischen Entwicklung. Leipzig, Engelmann, 1894.

186 Driesch H: Die organischen Regulationen. Leipzig, Engelmann, 1901.

187 Driever W, Nüsslein-Volhard C: The *bicoid* protein determines position in the *Drosophila* embryo in a concentration-dependent manner. Cell 1988;54:95–104.

188 Duboule D, Dollé P: The structural and functional organization of the murine HOX gene family resembles that of *Drosophila* homeotic genes. EMBO J 1989;8:1497–1505.

189 Duffy JB, Gergen JP: The *Drosophila* segmentation gene *runt* acts as a position-specific numerator element necessary for the uniform expression of the sex-determining gene *Sex-lethal*. Genes Dev 1991;5:2176–2187.

190 Duffy JB, Kania MA, Gergen JP: Expression and function of the *Drosophila* gene *runt* in early stages of neural development. Development 1991;113:1223–1230.

191 Dun RB: Growth of the mouse coat. VI. Distribution and number of vibrissae in the house mouse. Aust J Biol Sci 1958;2:95–105.

192 Easter SS Jr: Rules of retinotectal mapmaking. BioEssays 1986;5:158–162.

193 Easter SS, Jr, Purves D, Rakic P, Spitzer NC: The changing view of neural specificity. Science 1985;230:507–511.

194 Eberhard W: Computer simulation of orb-web construction. Am Zool 1969;9:229–238.

195 Eberhard WG: Behavioral characters for the higher classification of orb-weaving spiders. Evolution 1982;36:1067–1095.

196 Ede DA: Cell behaviour and embryonic development. Int J Neurosci 1972;3:165–174.

197 Edelman GM: Origins and mechanisms of specificity in clonal selection; in Edelman GM (ed): Cellular Selection and Regulation in the Immune Response. New York, Raven Press, 1974, pp 1–38.

198 Edelman GM: Cell adhesion molecules. Science 1983;219:450–457.

199 Edelman GM: Cell-adhesion molecules: A molecular basis for animal form. Sci Am 1984;250:118–129.

200 Edelmann GM: Cell-adhesion and the molecular processes of morphogenesis. Ann Rev Biochem 1985;54:135–169.

201 Edelman GM: Cell adhesion molecules in the regulation of animal form and tissue pattern. Ann Rev Cell Biol 1986;2: 81–116.

202 Edelman GM: Neural Darwinism: The Theory of Neuronal Group Selection. New York, Basic Books, 1987.

203 Edelman GM: Topobiology: An Introduction to Molecular Embryology. New York, Basic Books, 1988.

204 Edelman GM: Topobiology. Sci Am 1989;260:76–88.

205 Edelman GM, Gallin WJ: Cell adhesion as a basis of pattern in embryonic development. Am Zool 1987;27:645–656.

206 Edgar BA, Kiehle CP, Schubiger G: Cell cycle control by the nucleocytoplasmic ratio in early *Drosophila* development. Cell 1986;44:365–372.

207 Edgar BA, Odell GM, Schubiger G: A genetic switch, based on negative regulation, sharpens stripes in *Drosophila* embryos. Dev Genet 1989;10:124–142.

208 Edwards JL: Two perspectives on the evolution of the tetrapod limb. Am Zool 1989;29:235–254.

209 Eisen JS: Determination of primary motoneuron identity in developing zebrafish embryos. Science 1991;252:569–572.

210 Elder D: Theory of epigenetic coding. J Theor Biol 1984;108:327–332.

211 Ellinger MS: Ontogeny of melanophore patterns in haploid and diploid embryos of the frog, *Bombina* orientalis. J Morphol 1979;162:77–91.

212 Ellinger MS, Murphy JA: Cellular morphology in haploid amphibian embryos. J Embryol Exp Morphol 1980;59:249–261.
213 Ellis HM, Horvitz HR: Genetic control of programmed cell death in the nematode C. elegans. Cell 1986;44:817–829.
214 Ellis HM, Spann DR, Posakony JW: extramacrochaetae, a negative regulator of sensory organ development in Drosophila, defines a new class of helix-loop-helix proteins. Cell 1990;61:27–38.
215 Elsdale T, Pearson M: Somitogenesis in amphibia. II. Origins in early embryogenesis of two factors involved in somite specification. J Embryol Exp Morphol 1979;53: 245–267.
216 Elsdale T, Wasoff F: Fibroblast cultures and dermatoglyphics: the topology of two planar patterns. Roux Arch 1976;180:121–147.
217 Erickson CA, Oliver KR: Negative chemotaxis does not control quail neural crest dispersion. Dev Biol 1983;96:542–551.
218 Erickson JW, Cline TW: Molecular nature of the Drosophila sex determination signal and its link to neurogenesis. Science 1991;251:1071–1074.
219 Erneux T, Hiernaux J, Nicolis G: Turing's theory in morphogenesis. B Math Biol 1978;40:771–789.
220 Escher MC: Introduction; in: The Graphic Work of M.C. Escher. New York, Ballantine, 1960, pp 7–16.
221 Estruch JJ, Prinsen E, Onckelen HV, Schell J, Spena A: Viviparous leaves produced by somatic activation of an inactive cytokinin-synthesizing gene. Science 1991;254: 1364–1367.
222 Ettensohn CA: Mechanisms of epithelial invagination. Q Rev Biol 1985;60:289–307.
223 Evered D, Marsh J (ed): Cellular Basis of Morphogenesis. Ciba Fd Symp 1989; 144.
224 Fankhauser G: The effects of changes in chromosome number on amphibian development. Q Rev Biol 1945;20:20–78.
225 Fausto-Sterling A, Hsieh L: In vitro culture of Drosophila imaginal disc cells: Aggregation, sorting out, and differentiative abilities. Dev Biol 1987;120:284–293.
226 Fehon RG, Gauger A, Schubiger G: Cellular recognition and adhesion in embryos and imaginal discs of Drosophila melanogaster; in Loomis WF (ed): Genetic Regulation of Development. Symp Soc Dev Biol, vol 45. New York, Liss, 1987, pp 141–170.
227 Field RJ: Chemical organization in time and space. Am Sci 1985;73:142–150.
228 Finney M, Ruvkun G: The unc-86 gene product couples cell lineage and cell identity in C. elegans. Cell 1990;63:895–905.
229 Finney M, Ruvkun G, Horvitz HR: The C. elegans cell lineage and differentiation gene unc-86 encodes a protein with a homeodomain and extended similarity to transcription factors. Cell 1988;55:757–769.
230 Fletcher NH, Rossing TD: The Physics of Musical Instruments. New York, Springer, 1990.
231 Foerster P, Müller SC, Hess B: Curvature and spiral geometry in aggregation patterns of Dictyostelium discoideum. Development 1990;109:11–16.
232 Fortini ME, Rubin GM: Analysis of cis-acting requirements of the Rh3 and Rh4

genes reveals a bipartite organization to the rhodopsin promoters in *Drosophila melanogaster.* Genes Dev 1990;4:444–463.

233 Fowler JA: Control of vertebral number in teleosts – an embryological problem. Q Rev Biol 1970;45:148–167.

234 Frankel AD, Kim PS: Modular structure of transcription factors: Implications for gene regulation. Cell 1991;65:717–719.

235 Frankel J: Pattern Formation: Ciliate Studies and Models. New York, Oxford University Press, 1989.

236 Fraser SE: A differential adhesion approach to the patterning of nerve connections. Dev Biol 1980;79:453–464.

237 Fraser SE: Adhesive interactions and the patterning of nerve connections: an experimental and theoretical approach. Am Zool 1987;27:207–218.

238 Fraser SE: Multiple cues in neuronal patterning; in Malacinski GM (ed): Cytoplasmic Organization Systems. New York, McGraw-Hill, 1990, pp 155–174.

239 Fraser SE, Hunt RK: Retinotectal specificity: Models and experiments in search of a mapping function. Ann Rev Neurosci 1980;3:319–352.

240 French V: Pattern regulation and regeneration. Phil Trans R Soc Lond [B] 1981;295: 601–617.

241 French V: Development and evolution of the insect segment; in Goodwin BC, Holder N, Wylie CC (eds): Development and Evolution. Symp Brit Soc Dev Biol, vol 6. Cambridge, Cambridge University Press, 1983, pp 161–193.

242 French V: A model of insect limb regeneration; in Malacinski GM, Bryant SV (eds): Pattern Formation: A Primer in Developmental Biology. New York, Macmillan, 1984, pp 339–364.

243 French V: Gradients and insect segmentation. Development 1988;104(suppl):3–16.

244 French V: The development of segments in the invertebrates. Semin Dev Biol 1990; 1:89–100.

245 French V, Bryant PJ, Bryant SV: Pattern regulation in epimorphic fields. Science 1976;193:969–981.

246 Fristrom D: Cellular degeneration in the production of some mutant phenotypes in *Drosophila melanogaster.* Mol Gen Genet 1969;103:363–379.

247 Fristrom D: The cellular basis of epithelial morphogenesis: A review. Tiss Cell 1988; 20:645–690.

248 Garcia-Bellido A: Genetic control of wing disc development in *Drosophila;* in Porter R, Rivers J (eds): Cell Patterning. Ciba Found Symp, vol 29. Amsterdam, Elsevier, 1975, pp 161–182.

249 García-Bellido A: The cellular interphase; in Evered D, Marsh J (eds): Cellular Basis of Morphogenesis. Ciba Found Symp 1989;144:5–15.

250 García-Bellido A: Homoeotic and atavic mutations in insects. Am Zool 1977;17: 613–629.

251 García-Bellido A: Inductive mechanisms in the process of wing vein formation in *Drosophila.* Roux Arch 1977;182:93–106.

252 García-Bellido A, Ripoll P, Morata G: Developmental compartmentalisation of the wing disk of *Drosophila.* Nature New Biol 1973;245:251–253.

253 Gardiner DM, Bryant SV: Organization of positional information in the axolotl limb. J Exp Zool 1989;251:47–55.

254 Gardner M: The fantastic combinations of John Conway's new solitare game 'life'. Sci Am 1970;223:120–123.

255 Gardner M: On cellular automata, self-reproduction, the Garden of Eden and the game 'life'. Sci Am 1971;224:112–117.

256 Gardner M: Wheels, Life and Other Mathematical Amusements. New York, Freeman, 1983.

257 Garrell J, Campuzano S: The helix-loop-helix domain: A common motif for bristles, muscles, and sex. BioEssays 1991;13:493–498.

258 Garrell J, Modolell J: The *Drosophila extramacrochaetae* locus, an antagonist of proneural genes that, like these genes, encodes a helix-loop-helix protein. Cell 1990; 61:39–48.

259 Gaunt SJ: Expression patterns of mouse Hox genes: Clues to an understanding of developmental and evolutionary strategies. BioEssays 1991;13:505–513.

260 Gaunt SJ, Sharpe PT, Duboule D: Spatially restricted domains of homeo-gene transcripts in mouse embryos: Relation to a segmented body plan. Development 1988; 104(suppl):169–179.

261 Gaunt SJ, Singh PB: Homeogene expression patterns and chromosomal imprinting. Trends Genet 1990;6:208–212.

262 Gehring W: Übertragung und Änderung der Determinationsqualitäten in Antennenscheiben-Kulturen von *Drosophila melanogaster.* J Embryol Exp Morphol 1966;15: 77–111.

263 Gehring W: Clonal analysis of determination dynamics in cultures of imaginal disks in *Drosophila melanogaster.* Dev Biol 1967;16:438–456.

264 Gehring W: The stability of the determined state in cultures of imaginal disks in *Drosophila;* in Ursprung H, Nöthiger R (eds): The Biology of Imaginal Disks. Results and Problems in Cell Differentiation, vol 5. New York, Springer, 1972, pp 35–58.

265 Gehring WJ: Imaginal discs: determination; in Ashburner M, Wright TRF (eds): The Genetics and Biology of *Drosophila*, vol 2c. New York, Academic Press, 1978, pp 511–554.

266 Gehring WJ, Nöthiger R: The imaginal discs of *Drosophila;* in Counce SJ, Waddington CH (eds): Developmental Systems: Insects, vol 2. New York, Academic Press, 1973, pp 211–290.

267 Gelbart WM: The *decapentaplegic* gene: A TGF-β homologue controlling pattern formation in *Drosophila.* Development 1989;(suppl):65–74.

268 George FH: Cybernetics and Biology. San Francisco, Freeman, 1965.

269 Gergen JP, Coulter D, Wieschaus E: Segmental pattern and blastoderm cell identities; in Gall JG (ed): Gametogenesis and the Early Embryo. Symp Soc Dev Biol, vol 44. New York, Liss, 1986, pp 195–220.

270 Gerhardt M, Schuster H, Tyson JJ: A cellular automation model of excitable media including curvature and dispersion. Science 1990;247:1563–1566.

271 Gerhart J: Embryonic development: Toward a synthesis. Science 1989;243:1373–1374.

272 Gerhart J: The primacy of cell interactions in development. Trends Genet 1989;5: 233–236.

273 Gerhart J, Black S, Scharf S, Gimlich R, Vincent J-P, Danilchik M, Rowning B, Roberts J: Amphibian early development. BioScience 1986;36:541–549.

274 Gerhart JC: Mechanisms regulating pattern formation in the amphibian egg and early embryo; in Goldberger RF (ed): Biological Regulation and Development: Molecular Organization and Cell Function, vol 2. New York, Plenum Press, 1980, pp 133–316.

275 Ghysen A, Dambly-Chaudière C: From DNA to form: the *achaete-scute* complex. Genes Dev 1988;2:495–501.

276 Ghysen A, Dambly-Chaudière C: Genesis of the *Drosophila* peripheral nervous system. Trends Genet 1989;5:251–255.

277 Gierer A: Molecular models and combinatorial principles in cell differentiation and morphogenesis. Cold Spring Harbor Symp Quant Biol 1974;38:951–961.

278 Gierer A, Berking S, Bode H, David CN, Flick K, Hansmann G, Schaller H, Trenkner E: Regeneration of hydra from reaggregated cells. Nature New Biol 1972; 239:98–101.

279 Gierer A, Meinhardt H: A theory of biological pattern formation. Kybernetik 1972; 12:30–39.

280 Gierer A, Meinhardt H: Biological pattern formation involving lateral inhibition; in: Lectures on Mathematics in the Life Sciences, vol 7. Providence, American Mathematics Society, 1974, pp 163–183.

281 Gilbert SF: Cytoplasmic action in development. Q Rev Biol 1991;66:309–316.

282 Gilbert SF: Developmental Biology, ed 3. Sunderland, Sinauer, 1991.

283 Girton JR: Pattern triplications produced by a cell-lethal mutation in *Drosophila.* Dev Biol 1981;84:164–172.

284 Girton JR: Genetically induced abnormalities in *Drosophila:* Two or three patterns? Am Zool 1982;22:65–77.

285 Girton JR: Morphological and somatic clonal analyses of pattern triplications. Dev Biol 1983;99:202–209.

286 Girton JR, Bryant PJ: The use of cell lethal mutations in the study of *Drosophila* development. Dev Biol 1980;77:233–243.

287 Girton JR, Kumor AL: The role of cell death in the induction of pattern abnormalities in a cell-lethal mutation of *Drosophila.* Dev Genet 1985;5:93–102.

288 Gmitro JI, Scriven LE: A physicochemical basis for pattern and rhythm; in Waddington CH (ed): Towards a Theoretical Biology. II. Sketches. Edinburgh, Edinburgh University Press, 1969, pp 184–203.

289 Gober JW, Champer R, Reuter S, Shapiro L: Expression of positional information during cell differentiation in *Caulobacter.* Cell 1991;64:381–391.

290 Goel NS, Thompson RL: Movable finite automata (MFA): A new tool for computer modeling of living systems; in Langton CG (ed): Artificial Life. New York, Addison-Wesley, 1989, pp 317–340.

291 Goldbeter A, Caplan SR: Oscillatory enzymes. Ann Rev Biophys Bioeng 1976;5: 449–476.

292 Goldbeter A, Wolpert L: Covalent modification of proteins as a threshold mechanism in development. J Theor Biol 1990;142:243–250.

293 Goldschmidt R: Die quantitativen Grundlagen von Vererbung und Artbildung. Berlin, Springer, 1920.

294 Goldschmidt R: Physiological Genetics. New York, McGraw-Hill, 1938.

295 Goldschmidt R: The Material Basis of Evolution. New Haven, Yale University Press, 1940.

296 Goldsmith TH: Optimization, constraint, and history in the evolution of eyes. Q Rev Biol 1990;65:281–322.

297 Gomer RH, Firtel RA: Cell-autonomous determination of cell-type choice in *Dictyostelium* development by cell-cycle phase. Science 1987;237:758–762.

298 González-Reyes A, Urquia N, Gehring WJ, Struhl G, Morata G: Are cross-regulatory interactions between homoeotic genes functionally significant? Nature 1990; 344:78–80.

299 Goodfield J: Changing strategies: A comparison of reductionist attitudes in biological and medical research in the nineteenth and twentieth centuries; in Ayala FJ, Dobzhansky T (eds): Studies in the Philosophy of Biology. Reduction and Related Problems. Berkeley, University of California Press, 1974, pp 65–86.

300 Goodman CS: Landmarks and labels that help developing neurons find their way. BioScience 1984;34:300–307.

301 Goodman CS, Bastiani MJ: How embryonic nerve cells recognize one another. Sci Am 1984;251:58–66.

302 Goodman CS, Raper JA, Ho RK, Chang S: Pathfinding by neuronal growth cones in grasshopper embryos; in Subtelny S, Green PB (eds): Developmental Order: Its Origin and Regulation. Symp Soc Dev Biol, vol 40. New York, Liss, 1982, pp 275–316.

303 Goodwin B, Saunders P (eds): Theoretical Biology: Epigenetic and Evolutionary Order from Complex Systems. Edinburgh, Edinburgh University Press, 1989.

304 Goodwin BC: Temporal Organization in Cells: A Dynamic Theory of Cellular Control Processes. New York, Academic Press, 1963.

305 Goodwin BC: Analytical Physiology of Cells and Developing Organisms. New York, Academic Press, 1976.

306 Goodwin BC: Mechanics, fields and statistical mechanics in developmental biology. Proc Roy Soc Lond [B] 1977;199:407–414.

307 Goodwin BC: Developing organisms as self-organizing fields; in Antonelli PL (ed): Mathematical Essays on Growth and the Emergence of Form. Edmonton, University of Alberta Press, 1985, pp 185–200.

308 Goodwin BC: What are the causes of morphogenesis? BioEssays 1985;3:32–36.

309 Goodwin BC: Problems and prospects in morphogenesis. Experientia 1988;44:633–637.

310 Goodwin BC, Cohen MH: A phase-shift model for the spatial and temporal organization of developing systems. J Theor Biol 1969;25:49–107.

311 Goodwin BC, Kauffman SA: Bifurcations, harmonics and the four-colour wheel model of *Drosophila* development; in Goldbeter A (ed): Cell-to-Cell Signalling: From Experiments to Theoretical Models. New York, Academic Press, 1989, pp 213–227.

312 Goodwin BC, Trainor LEH: The ontogeny and phylogeny of the pentadactyl limb; in Goodwin BC, Holder N, Wylie CC (eds): Development and Evolution. Symp Br Soc Dev Biol, vol 6. Cambridge, Cambridge University Press, 1983, pp 75–98.

313 Gordon R: On stochastic growth and form. Proc Natl Acad Sci USA 1966;56:1497–1504.

314 Gottlieb DI, Glaser L: Cellular recognition during neural development. Ann Rev Neurosci 1980;3:303–318.

315 Gould SJ: Ontogeny and Phylogeny. Cambridge, Harvard University Press, 1977.

316 Gould SJ: The evolutionary biology of constraint. Proc Am Acad Arts Sci 1980;109: 39–52.

317 Gould SJ: Is a new and general theory of evolution emerging? Paleobiology 1980;6: 119–130.

318 Gould SJ: What color is a zebra? Nat Hist 1981;90:16–22.

319 Graham CF: The effect of cell size and DNA content on the cellular regulation of DNA synthesis in haploid and diploid embryos. Exp Cell Res 1966;43:13–19.

320 Green H, Thomas J: Pattern formation by cultured human epidermal cells: development of curved ridges resembling dermatoglyphics. Science 1978;200:1385–1388.

321 Green JBA, Cooke J: Induction, gradient models and the role of negative feedback in body pattern formation in the amphibian embryo. Semin Dev Biol 1991;2:95–106.

322 Green JBA, Smith JC: Graded changes in dose of a *Xenopus* activin A homologue elicit stepwise transitions in embryonic cell fate. Nature 1990;347:391–394.

323 Green JBA, Smith JC: Growth factors as morphogens: Do gradients and thresholds establish body plan? Trends Genet 1991;7:245–250.

324 Green PB: Plasticity in shoot development: A biophysical view. Symp Soc Exp Biol 1986;40:212–232.

325 Green PB: Inheritance of pattern: analysis from phenotype to gene. Am Zool 1987; 27:657–673.

326 Greenwald I: The genetic analysis of cell lineage in *Caenorhabditis elegans.* Phil Trans R Soc Lond [B] 1985;312:129–137.

327 Greenwald I: Cell-cell interactions that specify certain cell fates in *C. elegans* development. Trends Genet 1989;5:237–241.

328 Grier JW: Biology of Animal Behavior. St. Louis, Times Mirror/Mosby, 1984.

329 Grimes GW, Aufderheide KJ: Cellular Aspects of Pattern Formation: The Problem of Assembly. Monogr Dev Biol, vol 22. Basel, Karger, 1991.

330 Grünbaum B, Shephard GC: Tilings and Patterns. New York, Freeman, 1987.

331 Gubb D: Domains, compartments and determinative switches in *Drosophila* development. BioEssays 1985;2:27–31.

332 Gubb D: Pattern formation during development and the design of the adult cuticle in *Drosophila;* in Balls M, Bownes M (eds): Metamorphosis. Symp Br Soc Dev Biol, vol 8. Oxford, Clarendon Press, 1985, pp 88–107.

333 Guillery RW: Visual pathways in albinos. Sci Am 1974;230:44–54.

334 Gurdon JB: A community effect in animal development. Nature 1988;336:772–774.

335 Gustafson T, Wolpert L: The cellular basis of morphogenesis and sea urchin development. Int Rev Cytol 1963;15:139–214.

336 Gustafson T, Wolpert L: Cellular movement and contact in sea urchin morphogenesis. Biol Rev 1967;42:442–498.

337 Gutowitz H (ed): Cellular Automata: Theory and Experiment. Cambridge, MIT Press, 1991.

338 Haas HJ: On the epigenetic mechanisms of patterns in the insect integument. A reappraisal of older concepts. Int Rev Genet Exp Zool 1968;3:1–51.

339 Hadorn E: Problems of determination and transdetermination. Brookhaven Symp Biol 1965;18:148–161.

340 Hafen E, Basler K: Specification of cell fate in the developing eye of *Drosophila*. Development 1991;(suppl 1991):123–130.

341 Hall BK: Epigenetic control in development and evolution; in Goodwin BC, Holder N, Wylie CG (eds): Development and Evolution. Cambridge, Cambridge University Press, 1983, pp 353–379.

342 Hall PA, Watt FM: Stem cells: The generation and maintenance of cellular diversity. Development 1989;106:619–633.

343 Hamburger V: The Heritage of Experimental Embryology: Hans Spemann and the Organizer. New York, Oxford University Press, 1988.

344 Hameroff S, Rasmussen S, Mansson B: Molecular automata in microtubules: basic computational logic of the living state? In Langton CG (ed): Artificial Life. New York, Addison-Wesley, 1989, pp 521–553.

345 Hameroff SR: Ultimate Computing: Biomolecular Consciousness and Nanotechnology. Amsterdam, Elsevier, 1987.

346 Hamilton L: The formation of somites in *Xenopus*. J Embryol Exp Morphol 1969; 22:253–264.

347 Hanover JW: Control of tree growth. Bioscience 1980;30:756-762.

348 Hanscombe O, Whyatt D, Fraser P, Yannoutsos N, Greaves D, Dillon N, Grosveld F: Importance of globin gene order for correct developmental expression. Genes Dev 1991;5:1387–1394.

349 Haraway DJ: Crystals, Fabrics, and Fields: Metaphors of Organicism in Twentieth-Century Developmental Biology. New Haven, Yale University Press, 1976.

350 Harding K, Rushlow C, Doyle HJ, Hoey T, Levine M: Cross-regulatory interactions among pair-rule genes in *Drosophila*. Science 1986;233:953–959.

351 Harris AK, Stopak D, Warner P: Generation of spatially periodic patterns by a mechanical instability: a mechanical alternative to the Turing model. J Embryol Exp Morphol 1984;80:1–20.

352 Harrison LG: An overview of kinetic theory in developmental modeling; in Subtelny S, Green PB (eds): Developmental Order: Its Origin and Regulation. Symp Soc Dev Biol, vol 40. New York, Liss, 1982, pp 3–33.

353 Harrison LG, Tan KY: Where may reaction-diffusion mechanisms be operating in metameric patterning of *Drosophila* embryos? BioEssays 1988;8:118–124.

354 Hartenstein V, Campos-Ortega JA: Early neurogenesis in wild-type *Drosophila melanogaster*. Roux's Arch Dev Biol 1984;193:308–325.

355 Hartenstein V, Posakony JW: Development of adult sensilla on the wing and notum of *Drosophila melanogaster*. Development 1989;107:389–405.

356 Hartenstein V, Posakony JW: A dual function of the *Notch* gene in *Drosophila* sensillum development. Dev Biol 1990;142:13–30.

357 Haselkorn R: Heterocysts. Ann Rev Plant Physiol 1978;29:319–344.

358 Hassell MP, Comins HN, May RM: Spatial structure and chaos in insect population dynamics. Nature 1991;353:255–258.

359 Haughn GW, Somerville CR: Genetic control of morphogenesis in *Arabidopsis*. Dev Genet 1988;9:73–89.

360 Hayes B: The cellular automation offers a model of the world and a world unto itself. Sci Am 1984;250:12–21.

361 Hayes B: Turning turtle gives one a view of geometry from the inside out. Sci Am 1984;250:14–20.

362 Haynie JL: Homologies of positional information in thoracic imaginal discs of *Drosophila melanogaster.* Roux Arch 1982;191:293–300.

363 Hedgecock EM: Cell lineage mutants in the nematode *Caenorhabditis elegans.* Trends Neurosci 1985;8:288–293.

364 Hedges ES: Liesegang Rings and Other Periodic Structures. London, Chapman & Hall, 1932.

365 Heitzler P, Simpson P: The choice of cell fate in the epidermis of *Drosophila.* Cell 1991;64:1083–1092.

366 Held LI Jr: A high-resolution morphogenetic map of the second-leg basitarsus in *Drosophila melanogaster.* Roux Arch 1979;187:129–150.

367 Held LI Jr: Pattern as a function of cell number and cell size on the second-leg basitarsus of *Drosophila.* Roux Arch 1979;187:105–127.

368 Held LI Jr: Arrangement of bristles as a function of bristle number on a leg segment in *Drosophila melanogaster.* Roux Arch Dev Biol 1990;199:48–62.

369 Held LI Jr: Bristle patterning in *Drosophila.* BioEssays 1991;13:633–640.

370 Held LI Jr, Bryant PJ: Cell interactions controlling the formation of bristle patterns in *Drosophila;* in Malacinski GM, Bryant SV (eds): Pattern Formation: A Primer in Developmental Biology. New York, Macmillan, 1984, pp 291–322.

371 Held LI Jr, Duarte CM, Derakhshanian K: Extra tarsal joints and abnormal cuticular polarities in various mutants of *Drosophila melanogaster.* Roux Arch Dev Biol ö986;195:145–157.

372 Held LI Jr, Pham TT: Accuracy of bristle placement on a leg segment in *Drosophila melanogaster.* J Morphol 1983;178:105–110.

373 Henisch HK: Crystals in Gels and Liesegang Rings. New York, Cambridge University Press, 1988.

374 Herr W: An agent of suppression. Nature 1991;350:554–555.

375 Herskowitz I: Master regulatory loci in yeast and lambda. Cold Spring Harb Symp Quant Biol 1985;50:565–574.

376 Hidalgo A: Interactions between segment polarity genes and the generation of the segmental pattern in *Drosophila.* Mech Dev 1991;35:77–87.

377 Hidalgo A, Ingham P: Cell patterning in the *Drosophila* segment: Spatial regulation of the segment polarity gene *patched.* Development 1990;110:291–301.

378 Hinchliffe JR: Cell death in embryogenesis; in Bowen ID, Lockshin RA (eds): Cell Death in Biology and Pathology. London, Chapman & Hall, 1981, pp 35–78.

379 Hinchliffe JR, Johnson DR: The Development of the Vertebrate Limb: An Approach through Experiment, Genetics, and Evolution. Oxford, Clarendon Press, 1980.

380 Ho MW, Saunders PT: Beyond neo-Darwinism – an epigenetic approach to evolution. J Theor Biol 1979;78:573–591.

381 Hodges A: Alan Turing: The Enigma. New York, Simon & Schuster, 1983.

382 Hodgkin J: Sex determination compared in *Drosophila* and *Caenorhabditis.* Nature 1990;344:721–728.

383 Holder N: Developmental constraints and the evolution of vertebrate digit patterns. J Theor Biol 1983;104:451–471.

384 Holder N: The vertebrate limb: patterns and constraints in development and evolution; in Goodwin BC, Holder N, Wylie CC (eds): Development and Evolution. Symp

Br Soc Dev Biol, vol 6. Cambridge, Cambridge University Press, 1983, pp 399–425.

385 Holder N, Glade R: Skin glands in the axolotl: the creation and maintenance of a spacing pattern. J Embryol Exp Morphol 1984;79:97–112.

386 Holland PWH: Homeobox genes and segmentation: co-option, co-evolution, and convergence. Semin Dev Biol 1990;1:135–145.

387 Holland PWH, Hogan BLM: Expression of homeo box genes during mouse development: a review. Genes Dev 1988;2:773–782.

388 Holliday R: Mechanisms for the control of gene activity during development. Biol Rev 1990;65:431–471.

389 Hollingsworth MJ: Sex-combs of intersexes and the arrangement of the chaetae on the legs of Drosophila. J Morphol 1964;115:35–51.

390 Holtzer H: Proliferative and quantal cell cycles in the differentiation of muscle, cartilage, and red blood cells; in Padykula HA (ed): Control Mechanisms in the Expression of Cellular Phenotypes. Symp Int Soc Cell Biol, vol 9. New York, Academic Press, 1970, pp 69–88.

391 Holtzer H: Cell lineages, stem cells and the 'quantal' cell cycle concept; in Lord BI, Potten CS, Cole RJ (eds): Stem Cells and Tissue Homeostasis. Symp Br Soc Cell Biol, vol 2. Cambridge, Cambridge University Press, 1978, pp 1–27.

392 Holtzer H, Rubinstein N, Fellini S, Yeoh G, Chi J, Birnbaum J, Okayama M: Lineage, quantal cell cycles, and the generation of diversity. Q Rev Biophys 1975;8:523–557.

393 Hooper JE, Scott MP: The Drosophila patched gene encodes a putative membrane protein required for segmental patterning. Cell 1989;59:751–765.

394 Hopcroft JE: Turing machines. Sci Am 1984;250:86–98.

395 Hörstadius S: Experimental Embryology of Echinoderms. Oxford, Clarendon Press, 1973.

396 Horvitz HR, Sternberg PW: Multiple intercellular signalling systems control the development of the Caenorhabditis elegans vulva. Nature 1991;351:535–541.

397 Howard K: The generation of periodic pattern during early Drosophila embryogenesis. Development 1988;104(suppl):35–50.

398 Howard KR, Struhl G: Decoding positional information: Regulation of the pair-rule gene hairy. Development 1990;110:1223–1231.

399 Hronowski L, Gillespie LL, Armstrong JB: Development and survival of haploids of the Mexican axolotl, Ambystoma mexicanum. J Exp Zool 1979;209:41–48.

400 Huang F, Dambly-Chaudière C, Ghysen A: The emergence of sense organs in the wing disc of Drosophila. Development 1991;111:1087–1095.

401 Hubel DH, Wiesel TN, Le Vay S: Plasticity of ocular dominance columns in monkey striate cortex. Phil Trans R Soc Lond [B]1977;278:377–409.

402 Hülskamp M, Tautz D: Gap genes and gradients – the logic behind the gaps. BioEssays 1991;13:261–268.

403 Hunt P, Whiting J, Muchamore I, Marshall H, Krumlauf R: Homeobox genes and models for patterning the hindbrain and branchial arches. Development 1991;(suppl 1):187–196.

404 Hunter T: A thousand and one protein kinases. Cell 1987;50:823–829.

405 Huszagh VA, Infante JP: The hypothetical way of progress. Nature 1989;338:109.

406 Huxley J: A discussion on ritualization of behaviour in animals and man. Introduction. Phil Trans R Soc Lond [B] 1966;251:249–271.
407 Huxley JS: Problems of Relative Growth, ed. 2. London, Methuen, 1932.
408 Huxley JS, de Beer GR: The Elements of Experimental Embryology. Cambridge, Cambridge University Press, 1934.
409 Immerglück K, Lawrence PA, Bienz M: Induction across germ layers in Drosophila mediated by a genetic cascade. Cell 1990;62:261–268.
410 Ingham P: The molecular genetics of embryonic pattern formation in Drosophila. Nature 1988;335:25–34.
411 Ingham PW: Genetic control of segmental patterning in the Drosophila embryo; in Mahowald AP (ed): Genetics of Pattern Formation and Growth Control. Symp Soc Dev Biol, vol 48. New York, Wiley, 1990, pp 181–196.
412 Inoué S: The role of self-assembly in the generation of biological form; in Subtelny S, Green PB (eds): Developmental Order: Its Origin and Regulation. Symp Soc Dev Biol, vol 40. New York, Liss, 1982, pp 35–76.
413 Ish-Horowicz D, Gyurkovics H: Ectopic segmentation gene expression and metameric regulation in Drosophila. Development 1988;104(suppl):67–73
414 Ish-Horowicz D, Pinchin SM, Ingham PW, Gyrukovics HG: Autocatalytic ftz activation and metameric instability induced by ectopic ftz expression. Cell 1989;57:223–232.
415 Iten LE: Pattern specification and pattern regulation in the embryonic chick limb bud. Am Zool 1982;22:117–129.
416 Itow T: Inhibitors of DNA synthesis change the differentiation of body segments and increase the segment number in horseshoe crab embryos (Chelicerata, Arthropoda). Roux Arch Dev Biol 1986;195:323–333.
417 Jacob F: Evolution and tinkering. Science 1977;196:1161–1166.
418 Jacobson AG: Somitomeres: mesodermal segments of vertebrate embryos. Development 1988;(suppl)104:209–220.
419 Jacobson M: Segmentation and homeosis of vertebrate limbs. Semin Dev Biol 1990; 1:101–107.
420 James AA, Bryant PJ: Mutations causing pattern deficiencies and duplications in the imaginal wing disk of Drosophila melanogaster. Dev Biol 1981;85:39–54.
421 Javois LC: Pattern specification in the developing chick limb; in Malacinski GM, Bryant SV (eds): Pattern Formation: A Primer in Developmental Biology. New York, Macmillan, 1984, pp 557–579.
422 Jerne NK: The generative grammar of the immune system. Science 1985;229:1057–1059.
423 Ji S: Biocybernetics. A machine theory of biology; in Ji S (ed): Molecular Theories of Cell Life and Death. New Brunswick, Rutgers University Press, 1991, pp 1–237.
424 Jiang J, Hoey T, Levine M: Autoregulation of a segmentation gene in Drosophila: Combinatorial interaction of the even-skipped homeo box protein with a distal enhancer element. Genes Dev 1991;5:265–277.
425 Johnston NV: The pattern of abdominal microchaetae in Drosophila. Aust J Biol Sci 1966;19:155–166.
426 Jürgens H, Peitgen H-O, Saupe D: The language of fractals. Sci Am 1990;263:60–67.

427 Jursnich VA, Fraser SE, Held LI Jr, Ryerse J, Bryant PJ: Defective gap-junctional communication associated with imaginal disc overgrowth and degeneration caused by mutations of the *dco* gene in *Drosophila*. Dev Biol 1990;140:413–429.

428 Kafatos FC: Sequential cell polymorphism: A fundamental concept in developmental biology. Adv Insect Physiol 1976;12:1–15.

429 Kagan ML, Sachs T: Development of immature stomata: evidence for epigenetic selection of a spacing pattern. Dev Biol 1991;146:100–105.

430 Kalil RE: Synapse formation in the developing brain. Sci Am 1989;261:76–85.

431 Kargon R, Achinstein P (eds): Kelvin's Baltimore Lectures and Modern Theoretical Physics: Historical and Philosophical Perspectives. Cambridge, MIT Press, 1987.

432 Karlsson J: A major difference between transdetermination and homeosis. Nature 1979;279:426–428.

433 Katz MJ: Templeting and self-assembly. J Theor Biol 1985;113:1–13.

434 Katz MJ: Templets and the Explanation of Complex Patterns. Cambridge, Cambridge University Press, 1986.

435 Katz MJ, Grenander U: Developmental matching and the numerical matching hypothesis for neuronal cell death. J Theor Biol 1982;98:501–517.

436 Katz MJ, Lasek RJ: Evolution of the nervous system: role of ontogenetic mechanisms in the evolution of matching populations. Proc Nat Acad Sci USA 1978;75:1349–1352.

437 Katz MJ, Lasek RJ: Guidance cue patterns and cell migration in multicellular organisms. Cell Motility 1980;1:141–157.

438 Kauffman S: Control circuits for determination and transdetermination: interpreting positional information in a binary epigenetic code; in Porter R, Rivers J (eds): Cell Patterning. Ciba Found Symp, vol 29. Amsterdam, Elsevier, 1975, pp 201–221.

439 Kauffman SA: Gene regulation networks: a theory for their global structure and behaviors. Curr Top Dev Biol 1971;6:145–182.

440 Kauffman SA: Control circuits for determination and transdetermination. Science 1973;181:310–318.

441 Kauffman SA: Bifurcations in insect morphogenesis. Part I; in Enns RH, Jones BL, Miura RM, Rangnekar SS (eds): Nonlinear Phenomena in Physics and Biology. New York, Plenum Press, 1981, pp 401–450.

442 Kauffman SA: Bifurcations in insect morphogenesis. Part II; in Enns RH, Jones BL, Miura RM, Rangnekar SS (eds): Nonlinear Phenomena in Physics and Biology. New York, Plenum Press, 1981, pp 451–484.

443 Kauffman SA: Pattern formation in the *Drosophila* embryo. Phil Trans R Soc Lond [B] 1981;295:567–594.

444 Kauffman SA: Developmental constraints: internal factors in evolution; in Goodwin BC, Holder N, Wylie CC (eds): Development and Evolution. Symp Br Soc Dev Biol, vol 6. Cambridge, Cambridge University Press, 1983, pp 195–225.

445 Kauffman SA: Pattern generation and regeneration; in Malacinski GM, Bryant SV (eds): Pattern Formation: A Primer in Developmental Biology. New York, Macmillan, 1984, pp 73–102.

446 Kauffman SA: Developmental logic and its evolution. BioEssays 1987;6:82–87.

447 Kauffman SA: Antichaos and adaptation. Sci Am 1991;265:78–84.

448 Kauffman SA, Ling E: Regeneration by complementary wing disc fragments of *Drosophila melanogaster*. Dev Biol 1981;82:238–257.
449 Kauffman SA, Shymko RM, Trabert K: Control of sequential compartment formation in *Drosophila*. Science 1978;199:259–270.
450 Kaufman TC, Wakimoto BT: Genes that control high level developmental switches; in Bonner JT (ed): Evolution and Development. New York, Springer, 1982, pp 189–205.
451 Keil TA, Steinbrecht RA: Mechanosensitive and olfactory sensilla of insects; in King RC, Akai H (eds): Insect Ultrastructure, vol 2. New York, Plenum Press, 1984, pp 477–516.
452 Keller R: Cell rearrangement in morphogenesis. Zool Sci 1987;4:763–779.
453 Kellogg VL, Bell RG: Studies of variation in insects. Proc Wash Acad Sci 1904;6: 203–332.
454 Keynes R, Cook G: Cell-cell repulsion: Clues from the growth cone? Cell 1990;62: 609–610.
455 Keynes RJ, Stern CD: Mechanisms of vertebrate segmentation. Development 1988; 103:413–429.
456 Kiger JA Jr: The bithorax complex – a model for cell determination in *Drosophila*. J Theor Biol 1973;40:455–467.
457 Killackey HP: Pattern formation in the trigeminal system of the rat. Trends Neurosci 1980;3:303–305.
458 Kim SK, Kaiser D: Cell alignment required in differentiation of *Myxococcus xanthus*. Science 1990;249:926–928.
459 Kimble JE: Strategies for control of pattern formation in *Caenorhabditis elegans*. Phil Trans R Soc Lond [B] 1981;295:539–551.
460 Kimelman D, Kirschner M, Scherson T: The events of the midblastula transition in *Xenopus* are regulated by changes in the cell cycle. Cell 1987;48:399–407.
461 King J: Genetic control of organelle assembly at the molecular level. I. Introduction: From genes to organelles. Q Rev Biol 1980;55:329–333.
462 Kirk DL: The ontogeny and phylogeny of cellular differentiation in *Volvox*. Trends Genet 1988;4:32–36.
463 Kirk DL, Harper JF: Genetic, biochemical, and molecular approaches to *Volvox* development and evolution. Int Rev Cytol 1986;99:217–293.
464 Kirschner M, Newport J, Gerhart J: The timing of early developmental events in *Xenopus*. Trends Genet 1985;1:41–47.
465 Klingensmith J, Noll E, Perrimon N: The segment polarity phenotype of *Drosophila* involves differential tendencies toward transformation and cell death. Dev Biol 1989;134:130–145.
466 Kluge AG: The characterization of ontogeny; in Humphries CJ (ed): Ontogeny and Systematics. New York, Columbia University Press, 1988, pp 57–81.
467 Knuth DE: Algorithms. Sci Am 1977;236:63–80.
468 Koestler A: Beyond atomism and holism – the concept of the holon; in Koestler A, Smythies JR (eds): Beyond Reductionism. New Perspectives in the Life Sciences. New York, MacMillan, 1969, pp 192–227.
469 Komaki MK, Okada K, Nishino E, Shimura Y: Isolation and characterization of novel mutants of *Arabidopsis thaliana* defective in flower development. Development 1988;104:195–203.

470 Kopell N, Howard LN: Pattern formation in the Belousov reaction; in Lectures on Mathematics in the Life Sciences, vol 7. Providence, American Mathematics Society, 1974, pp 201–216.

471 Korn RW: Arrangement of stomata on the leaves of *Pelargonium zonale* and *Sedum stahlii.* Ann Bot 1972;36:325–333.

472 Korn RW: A neighboring-inhibition model for stomate patterning. Dev Biol 1981; 88:115–120.

473 Korn RW, Fredrick GW: Development of D-type stomata in the leaves of *Ilex crenata* var. *convexa.* Ann Bot 1973;37:647–656.

474 Kraut R, Levine M: Spatial regulation of the gap gene *giant* during *Drosophila* development. Development 1991;111:601–609.

475 Kühn A: Lectures on Developmental Physiology, ed 2. New York, Springer, 1971.

476 Kühn A, von Engelhardt M: Über die Determination des Symmetriesystems auf dem Vorderflügel von *Ephestia kühniella* Z. W. Roux Arch EntwMech Org 1933; 130:660–703.

477 Kuo CF, Xanthopoulos KG, Darnell JE Jr: Fetal and adult localization of C/EBP: Evidence for combinatorial action of transcription factors in cell specific gene expression. Development 1990;109:473–481.

478 Lacalli TC: Modeling the *Drosophila* pair-rule pattern by reaction-diffusion: gap input and pattern control in a 4-morphogen system. J Theor Biol 1990;144:171–194.

479 Lacalli TC, Harrison LG: From gradient to segments: Models for pattern formation in early *Drosophila* embryogenesis. Semin Dev Biol 1991;2:107–117.

480 Lacalli TC, Wilkinson DA, Harrison LG: Theoretical aspects of stripe formation in relation to *Drosophila* segmentation. Development 1988;104:105–113.

481 Langley JN: Note on regeneration of pre-ganglionic fibres of the sympathetic. J Physiol (Lond) 1895;18:280–284.

482 Langton CG (ed): Artificial Life. New York, Addison-Wesley, 1989.

483 Larison KD, Bremiller R: Early onset of phenotype and cell patterning in the embryonic zebrafish retina. Development 1990;109:567–576.

484 Larsen E, McLaughlin HMG: The morphogenetic alphabet: Lessons for simpleminded genes. BioEssays 1987;7:130–132.

485 Lawrence PA: Development and determination of hairs and bristles in the milkweed bug *Oncopeltus fasciatus* (Lygaeidae, Hemiptera). J Cell Sci 1966;1:475–498.

486 Lawrence PA: Gradients in the insect segment: The orientation of hairs in the milkweed bug *Oncopeltus fasciatus.* J Exp Biol 1966;44:607–620.

487 Lawrence PA: The development of spatial patterns in the integument of insects; in Counce SJ, Waddington CH (eds): Developmental Systems: Insects, vol 2. New York, Academic Press, 1973, pp 157–209.

488 Lawrence PA: Homoeotic selector genes – a working definition. BioEssays 1984;1: 227–229.

489 Lawrence PA: Molecular development: Is there a light burning in the hall? Cell 1985; 40:221.

490 Lawrence PA: Pair-rule genes: Do they paint stripes or draw lines? Cell 1987;51: 879–880.

491 Lawrence PA: The present status of the parasegment. Development 1988; 104(suppl):61–65.

492 Lawrence PA: Cell lineage and cell states in the *Drosophila* embryo; in Evered D, Marsh J (eds): Cellular Basis of Morphogenesis. Ciba Fnd Symp, vol 144. New York, Wiley, 1989, pp 131–149.

493 Lawrence PA, Struhl G, Morata G: Bristle patterns and compartment boundaries in the tarsi of *Drosophila.* J Embryol Exp Morphol 1979;51:195–208.

494 Lawrence PA, Wright DA: The regeneration of segment boundaries. Phil Trans R Soc Lond [B] 1981;295:595–599.

495 Le Gros Clark WE: Deformation patterns in the cerebral cortex; in Le Gros Clark WE, Medawar PB (eds): Essays on Growth and Form. Oxford, Clarendon, 1945, pp 1–22.

496 Lebovitz RM, Ready DF: Ommatidial development in *Drosophila* eye disc fragments. Dev Biol 1986;117:663–671.

497 Lechleiter J, Girard S, Peralta E, Clapham D: Spiral calcium wave propagation and annihilation in *Xenopus laevis* oocytes. Science 1991;252:123–126.

498 Lees AD, Waddington CH: The development of the bristles in normal and some mutant types of *Drosophila melanogaster.* Proc R Soc Lond [B] 1942;131:87–101.

499 Lehmann R, Frohnhöfer HG: Segmental polarity and identity in the abdomen of *Drosophila* is controlled by the relative position of gap gene expression. Development 1989;(suppl):21–29.

500 Lengyel I, Epstein IR: Modeling of Turing structures in the chlorite-iodide-malonic acid-starch reaction system. Science 1991;251:650–652.

501 Lettvin JY, Maturana HR, McCulloch WS, Pitts WH: What the frog's eye tells the frog's brain. Proc Inst Radio Eng 1959;47:1940–1951.

502 Leventhal AG, Vitek DJ, Creel DJ: Abnormal visual pathways in normally pigmented cats that are heterozygous for albinism. Science 1985;229:1395–1397.

503 Lewin B: Genes IV. New York, Oxford University Press, 1990.

504 Lewin R: Why is development so illogical? Science 1984;224:1327–1329.

505 Lewis EB: Genes and developmental pathways. Am Zool 1963;3:33–56.

506 Lewis EB: Genetic control and regulation of developmental pathways; in Locke M (ed): Role of Chromosomes in Development. Symp Soc Dev Biol, vol 23. New York, Academic Press, 1964, pp 231–252.

507 Lewis EB: A gene complex controlling segmentation in *Drosophila.* Nature 1978; 276:565–570.

508 Lewis J: Growth and determination in the developing limb; in Ede DA, Hinchliffe JR, Balls M (eds): Vertebrate Limb and Somite Morphogenesis. Cambridge, Cambridge University Press, 1977, pp 215–228.

509 Lewis J: Simpler rules for epimorphic regeneration: The polar-coordinate model without polar coordinates. J Theor Biol 1981;88:371–392.

510 Lewis J: Continuity and discontinuity in pattern formation; in Subtelny S, Green PB (eds): Developmental Order: Its Origin and Regulation. Symp Soc Dev Biol, vol 40. New York, Liss, 1982, pp 511–531.

511 Lewis J: Genes and segmentation. Nature 1989;341:382–383.

512 Lewis J, Slack JMW, Wolpert L: Thresholds in development. J Theor Biol 1977;65: 579–590.

513 Lewis, JH: Fate maps and the pattern of cell division: A calculation for the chick wing-bud. J Embryol Exp Morphol 1975;33:419–434.

514 Lewis JH: Rules for building the chick wing: discrete and continuous aspects of morphogenesis; in Lindenmayer A, Rozenberg G (eds): Automata, Languages, Development. New York, North-Holland, 1976, pp 97–108.

515 Lewis JH, Wolpert L: The principle of non-equivalence in development. J Theor Biol 1976;62:479–490.

516 Liddington RC, Yan Y, Moulai J, Sahli R, Benjamin TL, Harrison SC: Structure of simian virus 40 and 3.8-Å resolution. Nature 1991;354:278–284.

517 Lifschytz E: Uncoupling of gonial and spermatocyte stages by means of conditional lethal mutations in *Drosophila melanogaster.* Dev Biol 1978;66:571–578.

518 Lindenmayer A: Mathematical models for cellular interactions in development. I Filaments with one-sided inputs. J Theor Biol 1968;18:280–299.

519 Lindenmayer A: Mathematical models for cellular interactions in development. II. Simple and branching filaments with two-sided inputs. J Theor Biol 1968;18:300–315.

520 Lindenmayer A: Developmental algorithms: Lineage versus interactive control mechanisms; in Subtelny S, Green PB (eds): Developmental Order: Its Origin and Regulation. Symp Soc Dev Biol, vol 40. New York, Liss, 1982, pp 219–245.

521 Lindenmayer A: Positional and temporal control mechanisms in inflorescence development; in Barlow PW, Carr DJ (eds): Positional Controls in Plant Development, Cambridge, Cambridge University Press, 1984, pp 461–486.

522 Lindenmayer A, Prusinkiewicz P: Developmental models of multicellular organisms: a computer graphics perspective; in Langton CG (ed): Artificial Life. New York, Addison-Wesley, 1989, pp 221–249.

523 Lindsley DL, Grell EH: Genetic Variations of *Drosophila melanogaster.* Carnegie Inst Wash Publ, No 627, Washington, 1968.

524 Linsenmayer TF: Control of integumentary patterns in the chick. Dev Biol 1972;27:244–271.

525 Llinás RR: The intrinsic electrophysiological properties of mammalian neurons: insights into central nervous system function. Science 1988;242:1654–1664.

526 Locke M: What every epidermal cell knows; in Beament JWL, Treherne JE (eds): Insects and Physiology. Edinburgh, Oliver & Boyd, 1967, pp 69–82.

527 Locke M (ed): The Emergence of Order in Developing Systems. Symp Soc Dev Biol, vol 27. New York, Academic Press, 1968.

528 Locke M: Epidermal cells (Arthropoda); in Bereiter-Hahn J, Matoltsy AG, Richards KS (eds): Biology of the Integument, vol 1. Berlin, Springer, 1984, pp 502–522.

529 Locke M: Insect cells for the study of general problems in biology – somatic inheritance. Int J Insect Morphol Embryol 1988;17:419–436.

530 Lockshin RA: Cell death in metamorphosis; in Bowen ID, Lockshin RA (eds): Cell Death in Biology and Pathology, London, Chapman & Hall, 1981, pp 79–121.

531 Loer CM, Steeves JD, Goodman CS: Neuronal cell death in grasshopper embryos: variable patterns in different species, clutches, and clones. J Embryol Exp Morphol 1983;78:169–182.

532 Logan SK, Wensink PC: Ovarian follicle cell enhancers from the *Drosophila* yolk protein genes: different segments of one enhancer have different cell-type specificities that interact to give normal expression. Genes Dev 1990;4:613–623.

533 Loomis WF, White S, Dimond RL: A sequence of dependent stages in the development of *Dictyostelium discoideum.* Dev Biol 1976;53:171–177.

534 Lord BI, Potten CS, Cole RJ (ed): Stem Cells and Tissue Homeostasis. Symp Br Soc
     Cell Biol, vol 2. Cambridge, Cambridge University Press, 1978.
535 Lounibos LP: Initiation and maintenance of cocoon spinning behaviour by saturniid
     silkworms. Physiol Entomol 1976;1:195–206.
536 Løvtrup S: Epigenetics: A Treatise on Theoretical Biology. New York, Wiley,
     1974.
537 Løvtrup S: Epigenetic mechanisms in the early amphibian embryo: Cell differentia-
     tion and morphogenetic elements. Biol Rev 1983;58:91–130.
538 Lumsden AGS: Pattern formation in the molar dentition of the mouse. J Biol Bucc
     1979;7:77–103.
539 Lyons MJ, Harrison LG, Lakowski BC, Lacalli TC: Reaction diffusion modelling of
     biological pattern formation: Application to the embryogenesis of *Drosophila mela-
     nogaster.* Can J Phys 1990;68:772–777.
540 MacWilliams HK: A model of gradient interpretation based on morphogen binding.
     J Theor Biol 1978;72:385–411.
541 MacWilliams HK: Models of pattern formation in *Hydra* and *Dictyostelium.* Semin
     Dev Biol 1991;2:119–128.
542 Maden M, Gribbin MC, Summerbell D: Axial organisation in developing and regen-
     erating vertebrate limbs; in Goodwin BC, Holder N, Wylie CC (eds): Development
     and Evolution. Symp Br Soc Dev Biol, vol 6. Cambridge, Cambridge University
     Press, 1983, pp 381–397.
543 Maderson PFA: Embryonic tissue interactions as the basis for morphological change
     in evolution. Am Zool 1975;15:315–327.
544 Madore BF, Freedman WL: Computer simulations of the Belousov-Zhabotinsky
     reaction. Science 1983;222:437–438.
545 Madore BF, Freedman WL: Self-organizing structures. Am Sci 1987;75:252–259.
546 Magrassi L, Lawrence PA: The pattern of cell death in *fushi tarazu,* a segmentation
     gene of *Drosophila.* Development 1988;104:447–451.
547 Mahaffey JW, Kaufman TC: The homeotic genes of the Antennapedia Complex and
     the Bithorax Complex of *Drosophila;* in Malacinski GM (ed): Developmental Genet-
     ics of Higher Organisms. New York, Macmillan, 1987, pp 329–359.
548 Mahoney PA, Weber U, Onofrechuk P, Biessmann H, Bryant PJ, Goodman CS: The
     fat tumor suppressor gene in *Drosophila* encodes a novel member of the cadherin
     gene superfamily. Cell 1991;67:853–868.
549 Malacinski GM (ed): Cytoplasmic Organization Systems. New York, McGraw-Hill,
     1990.
550 Mandelbrot BB: The Fractal Geometry of Nature. New York, Freeman, 1977.
551 Markus M, Hess B: Isotropic cellular automation for modelling excitable media.
     Nature 1990;347:56–58.
552 Marr D: Vision. New York, Freeman, 1982.
553 Martindale MQ, Shankland M: Intrinsic segmental identity of segmental founder
     cells of the leech embryo. Nature 1990;347:672–674.
554 Martinez Arias A: A cellular basis for pattern formation in the insect epidermis.
     Trends Genet 1989;5:262–267.
555 Martinez Arias A, Baker NE, Ingham PW: Role of segment polarity genes in the
     definition and maintenance of cell states in the *Drosophila* embryo. Development
     1988;103:157–170.

556 Martinez-Arias A, Lawrence PA: Parasegments and compartments in the *Drosophila* embryo. Nature 1985;313:639–642.
557 Marx J: How embryos tell heads from tails. Science 1991;254:1586–1588.
558 Maynard Smith J: Continuous, quantized and modal variation. Proc R Soc Lond [B] 1960;152:397–409.
559 Maynard Smith J: The counting problem; in Waddington CH (ed): Towards a Theoretical Biology. I. Prolegomena. Chicago, Aldine, 1968, pp 120–124.
560 Maynard Smith J, Burian R, Kauffman S, Alberch P, Campbell J, Goodwin B, Lande R, Raup D, Wolpert L: Developmental constraints and evolution. Q Rev Biol 1985;60:265–287.
561 Maynard-Smith J, Sondhi KC: The arrangement of bristle in *Drosophila.* J Embryol Exp Morphol 1961,9:661–672.
562 Mayr E: Behavior programs and evolutionary strategies. Am Sci 1974;62:650–659.
563 Mayr E: How biology differs from the physical sciences; in Depew DJ, Weber BH (eds): Evolution at a Crossroads: The New Biology and the New Philosophy of Science. Cambridge, MIT Press, 1985, pp 43–63.
564 McConnell SK, Kaznowski CE: Cell cycle dependence of laminar determination in developing neocortex. Science 1991;254:282–285.
565 McKearin DM, Spradling AC: *bag-of-marbles:* A *Drosophila* gene required to initiate both male and female gametogenesis. Genes Dev 1990;4:2242–2251.
566 McMahon D: A cell-contact model for cellular position determination in development. Proc Natl Acad Sci USA 1973;70:2396–2400.
567 McNally JG, Cox EC: Geometry and spatial patterns in *Polysphondylium pallidum.* Dev Genet 1988;9:663–672.
568 McNally JG, Cox EC: Spots and stripes: The patterning spectrum in the cellular slime mould *Polysphondylium pallidum.* Development 1989;105:323–333.
569 Meinertzhagen IA: The development of neuronal connection patterns in the visual systems of insects; in Porter R, Rivers J (eds): Cell Patterning. Ciba Fd Symp, vol 29. Amsterdam, Elsevier, 1975, pp 265–288.
570 Meinhardt H: Morphogenesis of lines and nets. Differentiation 1976;6:117–123.
571 Meinhardt H: Models for the ontogenetic development of higher organisms. Rev Physiol Biochem Pharmacol 1978;80:47–104.
572 Meinhardt H: Space-dependent cell determination under the control of a morphogen gradient. J Theor Biol 1978;74:307–321.
573 Meinhardt H: Models of Biological Pattern Formation. New York, Academic Press, 1982.
574 Meinhardt H: The role of compartmentalization in the activation of particular control genes and in the generation of proximo-distal positional information in appendages. Am Zool 1982;22:209–220.
575 Meinhardt H: Cell determination boundaries as organizing regions for secondary embryonic fields. Dev Biol 1983;1983:375–385.
576 Meinhardt H: The threefold subdivision of segments and the initiation of legs and wings in insects. Trends Genet 1986;2:36–41.
577 Meinhardt H: Pattern formation and the activation of particular genes; in Goldbeter A (ed): Cell to Cell Signalling: From Experiments to Theoretical Models. New York, Academic Press, 1989, pp 189–212.

578 Meinhardt H: Determination borders as organizing regions in the generation of secondary embryonic fields: The intiation of legs and wings. Semin Dev Biol 1991;2: 129–138.

579 Meinhardt H, Gierer A: Applications of a theory of biological pattern formation based on lateral inhibition. J Cell Sci 1974;15:321–346.

580 Meinhardt H, Gierer A: Generation and regeneration of sequence of structures during morphogenesis. J Theor Biol 1980;85:429–450.

581 Meinhardt H, Gierer A: Generation of spatial sequences of structures during development of higher organisms; in: Lectures on Mathematics in the Life Sciences, vol 14. Providence, American Mathematics Society, 1981, pp 1–20.

582 Meister M, Wong ROL, Baylor DA, Shatz CJ: Synchronous bursts of action potentials in ganglion cells of the developing mammalian retina. Science 1991;252:939–943.

583 Melton DA: Pattern formation in animal development. Science 1991;252:234–241.

584 Meyer T: Cell signaling by second messenger waves. Cell 1991;64:675–678.

585 Meyerowitz EM, Bowman JL, Brockman LL, Drews GN, Jack T, Sieburth LE, Weigel D: A genetic and molecular model for flower development in *Arabidopsis thaliana.* Development 1991;(suppl 1):157–167.

586 Meyerowitz EM, Kankel DR: A genetic analysis of visual system development in *Drosophila melanogaster.* Dev Biol 1978;62:112–142.

587 Meyerowitz EM, Smyth DR, Bowman JL: Abnormal flowers and pattern formation in floral development. Development 1989;106:209–217.

588 Michaelson J: Cell selection in development. Biol Rev 1987;62:115–139.

589 Miller KD, Keller JB, Stryker MP: Ocular dominance column development: Analysis and simulation. Science 1989;245:605–616.

590 Minsky M (ed): Semantic Information Processing. Cambridge, MIT Press, 1968.

591 Minsky M, Papert S: Perceptrons. Cambridge, MIT Press, 1969.

592 Mirabito PM, Adams TH, Timberlake WE: Interactions of three sequentially expressed genes control temporal and spatial specificity in *Aspergillus* development. Cell 1989;57:859–868.

593 Mitchison GJ: Phyllotaxis and the Fibonacci series. Science 1977;196:270–275.

594 Mitchison GJ, Wilcox M: Rule governing cell division in *Anabaena.* Nature 1972; 239:110–111.

595 Mitchison GJ, Wilcox M: Alteration in heterocyst pattern of *Anabaena* produced by 7-azatryptophan. Nature 1973;246:229–233.

596 Mitchison GJ, Wilcox M, Smith RJ: Measurement of an inhibitory zone. Science 1976;191:866–868.

597 Mittenthal JE: The rule of normal neighbors: A hypothesis for morphogenetic pattern regulation. Dev Biol 1981;88:15–26.

598 Mlodzik M, Baker NE, Rbin GM: Isolation and expression of *scabrous,* a gene regulating neurogenesis in *Drosophila.* Genes Dev 1990;4:1848–1861.

599 Mlodzik M, Hiromi Y, Weber U, Goodman CS, Rubin GM: The *Drosophila seven-up* gene, a member of the steroid receptor gene superfamily, controls photoreceptor cell fate. Cell 1990;60:211–224.

600 Monk PB, Othmer HG: Relay, oscillations and wave propagation in a model of

Dictyostelium discoideum; in: Lectures on Mathematics in the Life Sciences, vol 21. Providence, American Mathematics Society, 1989, pp 87–122.

601 Montalenti G: A physiological analysis of the barred pattern in Plymouth Rock feathers. J Exp Zool 1934;69:269–345.

602 Moore JA: Science as a way of knowing – developmental biology. Am Zool 1987;27: 415–573.

603 Morata G, Garcia-Bellido A: Developmental analysis of some mutants of the bithorax system of *Drosophila.* Roux Arch 1976;179:125–143.

604 Morata G, Lawrence PA: Anterior and posterior compartments in the head of *Drosophila.* Nature 1978;274:473–474.

605 Morgan MJ: Power of the scientific metaphor. Nature 1978;276:125–126.

606 Morgan NG: Cell Signalling. New York, Guilford Press, 1989.

607 Morrison P: Termites and telescopes. NOVA (TV Program Transcript), 1979, pp 1–17.

608 Müller GB: Experimental strategies in evolutionary embryology. Am Zool 1991;31: 605–615.

609 Muller HJ: Further studies on the nature and causes of gene mutations. Proc 6th Int Congr Genet 1932;1:213–255.

610 Müller SC, Hess B: Spiral order in chemical reactions; in Goldbeter A (ed): Cell to Cell Signalling: From Experiments to Theoretical Models. New York, Academic Press, 1989, pp 503–520.

611 Müller SC, Kai S, Ross J: Curiosities in periodic precipitation patterns. Science 1982;216:635–637.

612 Muneoka K, Bryant S: Regeneration and development of vertebrate appendages. Symp Zool Soc Lond 1984;52:177–196.

613 Muneoka K, Bryant SV: Evidence that patterning mechanisms in developing and regenerating limbs are the same. Nature 1982;298:369–371.

614 Muneoka K, Bryant SV: Cellular contribution to supernumerary limbs resulting from the interaction between developing and regenerating tissues in the axolotl. Dev Biol 1984;105:179–187.

615 Murray J: Parameter space for Turing instabilities in reaction-diffusion mechanisms: A comparison of models. J Theor Biol 1982;98:143–163.

616 Murray JD: On pattern formation mechanisms for lepidopteran wing patterns and mammalian coat markings. Phil Trans R Soc Lond [B] 1981;295:473–496.

617 Murray JD: A pre-pattern formation mechanism for animal coat markings. J Theor Biol 1981;88:161–199.

618 Murray JD: How the leopard gets its spots. Sci Am 1988;258:80–87.

619 Murray JD: Mathematical Biology. New York, Springer, 1989.

620 Murray JD, Deeming DC, Ferguson MWJ: Size-dependent pigmentation-pattern formation in embryos of *Alligator mississippiensis:* Time of initiation of pattern generation mechanism. Proc R Soc Lond [B] 1990;239:279–293.

621 Murray JD, Maini PK: Pattern formation mechanisms – a comparison of reaction-diffusion and mechanochemical models; in Goldbeter A (ed): Cell to Cell Signalling: From Experiments to Theoretical Models. New York, Academic Press, 1989, pp 159–170.

622 Murray JD, Myerscough MR: Pigmentation pattern formation on snakes. J Theor Biol 1991;149:339–360.

623   Nagorcka BN: The role of a reaction-diffusion system in the initiation of skin organ primordia. I. The first wave of initiation. J Theor Biol 1986;121:449–475.

624   Nagorcka BN: A pattern formation mechanism to control spatial organization in the embryo of *Drosophila melanogaster.* J Theor Biol 1988;132:277–306.

625   Nagorcka BN: Wavelike isomorphic prepatterns in development. J Theor Biol 1989; 137:127–162.

626   Nagorcka BN, Mooney JR: The role of a reaction-diffusion system in the initiation of primary hair follicles. J Theor Biol 1985;114:243–272.

627   Napier JR: The human hand. Carolina Biol. Readers 1976;61:1–16.

628   Nardi JB: Epithelial invagination: Adhesive properties of cells can govern position and directionality of epithelial folding. Differentiation 1981;20:97–103.

629   Nardi JB: Induction of invagination in insect epithelium: paradigm for embryonic invagination. Science 1981;214:564–566.

630   Nardi JB: Neuronal pathfinding in developing wings of the moth *Manduca sexta.* Dev Biol 1983;95:163–174.

631   Nardi JB, Kafatos FC: Polarity and gradients in lepidopteran wing epidermis. I. Changes in graft polarity, form, and cell density accompanying transpositions and reorientations. J Embryol Exp Morphol 1976;36:469–487.

632   Nardi JB, Kafatos FC: Polarity and gradients in lepidopteran wing epidermis. II. The differential adhesiveness model: Gradient of a non-diffusible cell surface parameter. J Embryol Exp Morphol 1976;36:489–512.

633   Nardi JB, Magee-Adams SM: Formation of scale spacing patterns in a moth wing. I. Epithelial feet may mediate cell rearrangement. Dev Biol 1986;116:265–277.

634   Nardi JB, Stocum DL: Surface properties of regenerating limb cells: Evidence for gradation along the proximodistal axis. Differentiation 1983;25:27–31.

635   Newell PC: How cells communicate: The system used by slime moulds. Endeavour New Ser 1977;1:63–68.

636   Newman SA: Lineage and pattern in the developing vertebrate limb. Trends Genet 1988;4:329–332.

637   Newman SA, Comper WD: 'Generic' physical mechanisms of morphogenesis and pattern formation. Development 1990;110:1–18.

638   Newman SA, Frisch HL: Dynamics of skeletal pattern formation in developing chick limb. Science 1979;205:662–668.

329   Newport JW, Kirschner MW: A major developmental transition in early *Xenopus* embryos. I. Characterization and timing of cellular changes at midblastula stage. Cell 1982;30:675–686.

640   Newport JW, Kirschner MW: A major developmental transition in early *Xenopus* embryos. II. Control of the onset of transcription. Cell 1982;30:687–696.

641   Nickerson M: An experimental analysis of barred pattern formation in feathers. J Exp Zool 1944;95:361–397.

642   Nicolis G, Erneux T, Herschkowitz-Kaufman M: Pattern formation in reacting and diffusing systems. Adv Chem Phys 1978;38:263–315.

643   Nicolis G, Prigogine I: Self-Organization in Nonequilibrium Systems. New York, Wiley, 1977.

644   Nieuwkoop PD, Johnen AG, Albers B: The Epigenetic Nature of Early Chordate Development. Developmental and Cell Biology, vol 16. Cambridge, Cambridge University Press, 1985.

645 Nijhout HF: Metaphors and the role of genes in development. BioEssays 1990;12: 441–446.
646 Nijhout HF: The Development and Evolution of Butterfly Wing Patterns. Washington, Smithsonian Press, 1991.
647 Nübler-Jung K: Tissue polarity in an insect segment: Denticle patterns resemble spontaneously forming fibroblast patterns. Development 1987;100:171–177.
648 Nübler-Jung K, Mardini B: Insect epidermis: polarity patterns after grafting result from divergent adhesions between host and graft tissue. Development 1990;110: 1071–1079.
649 Nüsslein-Volhard C: Determination of the embryonic axes of Drosophila. Development 1991;(Suppl 1):1–10.
650 O'Shea PS: Physical fields and cellular organisation: field-dependent mechanisms of morphogenesis. Experientia 1988;44:684–694.
651 Odell GM, Oster G, Alberch P, Burnside B: The mechanical basis of morphogenesis. I. Epithelial folding and invagination. Dev Biol 1981;85:446–462.
652 Olson EN: MyoD family: A paradigm for development? Genes Dev 1990;4:1454–1461.
653 Oppenheim RW: Neuronal cell death and some related regressive phenomena during neurogenesis: A selective historical review and progress report; in Cowan WM (ed): Studies in Developmental Neurobiology. Essays in Honor of Viktor Hamburger. New York, Oxford University Press, 1981, pp 74–133.
654 Oppenheim RW: Cell death during development of the nervous system. Ann Rev Neurosci 1991;14:453–501.
655 Oppenheimer JM: Essays in the History of Embryology and Biology. Cambridge, MIT Press, 1967.
656 Oppenheimer P: The artificial menagerie; in Langton CG (ed): Artificial Life. New York, Addison-Wesley, 1989, pp 251–274.
657 Osborn JW: On the biological improbability of Zahnreihen as embryological units. Evolution 1973;26:601–607.
658 Osborn JW: On the control of tooth replacement in reptiles and its relationship to growth. J Theor Biol 1974;46:509–527.
659 Osborn JW: Morphogenetic gradients: fields versus clones; in Butler PM, Joysey KA (ed): Development, Function, and Evolution of Teeth. New York, Academic Press, 1978, pp 171–201.
660 Oster G: Lateral inhibition models of developmental processes. Math Biosci 1988; 90:265–286.
661 Oster G, Weliky M: Pattern and morphogenesis. Semin Dev Biol 1991;2:139–150.
662 Oster GF, Murray JD, Harris AK: Mechanical aspects of mesenchymal morphogenesis. J Embryol Exp Morphol 1983;78:83–125.
663 Oster GF, Shubin N, Murray JD, Alberch P: Evolution and morphogenetic rules: The shape of the vertebrate limb in ontogeny and phylogeny. Evolution 1988;42: 862–884.
664 Othmer HG: Current problems in pattern formation; in: Lectures on Mathematics in the Life Sciences, vol 9. Providence, American Mathematics Society, 1977, pp 57–85.
665 Othmer HG, Pate E: Scale-invariance in reaction-diffusion models of spatial pattern formation. Proc Natl Acad Sci USA 1980;77:4180–4184.

666 Ouweneel WJ: Developmental genetics of homeosis. Adv Genet 1976;18:179–248.

667 Ouyang Q, Swinney HL: Transition from a uniform state to hexagonal and striped Turing patterns. Nature 1991;352:610–612.

668 Pantkratz MJ, Gaul U, Hoch M, Seifert E, Nauber U, Gerwin N, Rothe M, Brönner G, Forsbach V, Goerlich K, Jäckle H: Overlapping gap gene activities generate pair-rule stripes and delimit the expression domains of homeotic genes along the longitudinal axis of the *Drosophila* blastoderm embryo; in Mahowald AP (ed): Genetics of Pattern Formation and Growth Control. Symp Soc Dev Biol, vol 48. New York, Wiley, 1990, pp 17–29.

669 Pankratz MJ, Jäckle H: Making stripes in the *Drosophila* embryo. Trends Genet 1990;6:287–292.

670 Papageorgiou S: A morphogen gradient model for pattern regulation. I. Formation of non-repetitive and repetitive structures. Biophys Chem 1980;11:183–190.

671 Papageorgiou S: A reaction-diffusion theory of morphogenesis with inherent pattern invariance under scale variations. J Theor Biol 1983;100:57–79.

672 Papert S: Mindstorms: Children, Computers, and Powerful Ideas. New York, Basic Books, 1980.

673 Parascandolo S, Abernathy A: MacUser guide to graphics formats. MacUser 1990;6: 266–276.

674 Parkhurst SM, Bopp D, Ish-Horowicz D: X: A ratio, the primary sex-determining signal in *Drosophila*, is transduced by helix-loop-helix proteins. Cell 1990;63:1179–1191.

675 Parkhurst SM, Ish-Horowicz D: Mis-regulating segmentation gene expression in *Drosophila*. Development 1991;111:1121–1135.

676 Pascal E, Tijan R: Different activation domains of Sp1 govern formation of multimers and mediate transcriptional synergism. Genes Dev 1991;5:1646–1656.

677 Patel NH, Schafer B, Goodman CS, Holmgren R: The role of segment polarity genes during *Drosophila* neurogenesis. Genes Dev 1989;3:890–904.

678 Pattee HH: How does a molecule become a message? in Lang A (ed): Communication in Development. Symp Soc Dev Biol, vol 28. New York, Academic Press, 1969, pp 1–16.

679 Pearson M, Elsdale T: Somitogenesis in amphibian embryos. I. Experimental evidence for an interaction between two temporal factors in the specification of somite pattern. J Embryol Exp Morphol 1979;51:27–50.

680 Penrose LS, Penrose R: Impossible objects: A special type of visual illusion. Br J Psych 1958;49:31–33.

681 Perelson AS, Maini PK, Murray JD, Oster GF: Nonlinear pattern selection in a mechanical model for morphogenesis. Math Biol 1986;24:525–541.

682 Peters PJ: Orb web construction: Interaction of spider (*Araneus diadematus* Cl.) and thread configuration. Anim Behav 1970;18:478–484.

683 Pfannenstiel HD: The ventral nerve cord signals positional information during segment formation in an annelid (*Ophryotrocha puerilis*, Polychaeta). Roux Arch Dev Biol 1984;194:32–36.

684 Pickett-Heaps JD, Northcote DH: Cell division in the formation of the stomatal complex of the young leaves of wheat. J Cell Sci 1966;1:121–128.

685 Plickert G: Low-molecular-weight factors from colonial hydroids affect pattern formation. Roux Arch Dev Biol 1987;196:248–256.

686 Plickert G, Heringer A, Hiller B: Analysis of spacing in a periodic pattern. Dev Biol 1987;120:399–411.

687 Poethig RS: Phase change and the regulation of shoot morphogenesis in plants. Science 1990;250:923–930.

688 Poodry CA: Epidermis: Morphology and development; in Ashburner M, Wright TRF (eds): The Genetics and Biology of Drosophila, vol 2d. New York, Academic Press, 1980, pp 443–497.

689 Pool R: Did Turing discover how the leopard got its spots? Science 1991;251:627.

690 Postlethwait JH: Pattern formation in the wing and haltere imaginal discs after irradiation of Drosophila melanogaster first instar larvae. Roux Arch 1975;178:29–50.

691 Postlethwait JH, Girton J: Development of antennal-leg homoeotic mutants in Drosophila melanogaster. Genetics 1974;76:767–774.

692 Postlethwait JH, Schneiderman HA: A clonal analysis of determination in Antennapedia, a homoeotic mutant of Drosophila melanogaster. Proc Nat Acad Sci USA 1969;64:176–183.

693 Postlethwait JH, Schneiderman HA: Pattern formation and determination in the antenna of the homoeotic mutant Antennapedia of Drosophila melanogaster. Dev Biol 1971;25:606–640.

694 Postlethwait JH, Schneiderman HA: Pattern formation in imaginal discs of Drosophila melanogaster after irradiation of embryos and young larvae. Dev Biol 1973;32:345–360.

695 Potten CS, Loeffler M: Stem cells: attributes, cycles, spirals, pitfalls and uncertainties. Lessons for and from the crypt. Development 1990;110:1001–1020.

696 Primmett DRN, Norris WE, Carlson GJ, Keynes RJ, Stern CD: Periodic segmental anomalies induced by heat shock in the chick embryo are associated with the cell cycle. Development 1989;105:119–130.

697 Primmett DRN, Stern CD, Keynes RJ: Heat shock causes repeated segmental anomalies in the chick embryo. Development 1988;104:331–339.

698 Pruitt RE: Molecular genetics of floral development in Arabidopsis thaliana. BioEssays 1991;13:347–349.

699 Prusinkiewicz P, Lindenmayer A: The Algorithmic Beauty of Plants. New York, Springer, 1990.

700 Purves D: Selective formation of synapses in the peripheral nervous system and the chemoaffinity hypothesis of neural specificity; in Cowan WM (ed): Studies in Developmental Neurobiology. Essays in Honor of Viktor Hamburger. New York, Oxford University Press, 1981, pp 231–242.

701 Purves D: A Trophic Theory of Neural Connections. Cambridge, Harvard Press, 1988.

702 Purves D, Lichtman JW: Principles of Neural Development. Sunderland, Sinauer, 1985.

703 Pye K, Tsoar H: Aeolian Sand and Sand Dunes. London, Unwin Hyman, 1990.

704 Raff RA, Kaufman TC: Embryos, Genes, and Evolution: The Developmental-Genetic Basis of Evolutionary Change. New York, Macmillan, 1983.

705 Ransick A: Reproductive cell specification during Volvox obversus development. Dev Biol 1991;143:185–198.

706 Ransom R: Computers and Embryos: Models in Developmental Biology. New York, Wiley, 1981.

707 Raper KB: The Dictyostelids. Princeton, Princeton University Press, 1984.

708 Rasmussen S, Karampurwala H, Vaidyanath R, Jensen KS, Hameroff S: Computational connectionism within neurons: A model of cytoskeletal automata subserving neural networks. Physica D 1990;42:428–449.

709 Raup DM: The geometry of coiling in gastropods. Proc Natl Acad Sci USA 1961;47:602–609.

710 Raup DM: Computer as aid in describing form in gastropod shells. Science 1962;138:150–152.

711 Raup DM: Geometric analysis of shell coiling: General problems. J Paleontol 1966;40:1178–1190.

712 Raup DM, Michelson A: Theoretical morphology of the coiled shell. Science 1965;147:1294–1295.

713 Ready DF: A multifaceted approach to neural development. Trends Neurosci 1989;12:102–110.

714 Ready DF, Hanson TE, Benzer S: Development of the *Drosophila* retina, a neurocrystalline lattice. Dev Biol 1976;53:217–240.

715 Ready DF, Tomlinson A, Lebovitz RM: Building an ommatidium: Geometry and genes; in Hilfer SR, Sheffield JB (eds): Development of Order in the Visual System. Berlin, Springer, 1986, pp 97–125.

716 Reed CF: Cues in the web-building process. Am Zool 1969;9:211–221.

717 Reif WE: Development of dentition and dermal skeleton in embryonic *Scyliorhinus canicula.* J Morphol 1980;166:275–288.

718 Reif W-E: A model of morphogenetic processes in the dermal skeleton of elasmobranchs. N Jb Geol Paläont Abh 1980;159:339–359.

719 Reinert J, Holtzer H (ed): Cell Cycle and Cell Differentation. Results and Problems in Cell Differentiation, vol 7. New York, Springer, 1975.

720 Richelle J, Ghysen A: Determination of sensory bristles and pattern formation in *Drosophila.* I. A model. Dev Biol 1979;70:418–437.

721 Richter PH, Schranner R: Leaf arrangement. Geometry, morphogenesis, classification. Naturwissenschaften 1978;65:319–327.

722 Riddihough G, Ish-Horowicz D: Individual stripe regulatory elements in the *Drosophila hairy* promoter respond to maternal, gap, and pair-rule genes. Genes Dev 1991;5:840–854.

723 Riedl R: Order in Living Organisms: A System Analysis of Evolution. New York, Wiley, 1978.

724 Rinzel J: Simple model equations for active nerve conduction and passive neuronal integration; in: Lectures on Mathematics in the Life Sciences, vol 8. Providence, American Mathematics Society, 1976, pp 125–164.

725 Ripoll P, El Messal M, Laran E, Simpson P: A gradient of affinities for sensory bristles across the wing blade of *Drosophila melanogaster.* Development 1988;103:757–767.

726 Rivlin R: The Algorithmic Image: Graphic Visions of the Computer Age. Richmond, Microsoft Press, 1986.

727 Roberts P: Mosaics involving aristapedia, a homeotic mutant of *Drosophila melanogaster.* Genetics 1964;49:593–598.

728 Robertson A: Variation in scutellar bristle number. An alternative hypothesis. Am Nat 1965;99:19–23.

729 Robertson A, Cohen MH: Quantitative analysis of the development of cellular slime molds. II. Lectures on Mathematics in the Life Sciences, vol 6. Providence, American Mathematics Society, 1974, pp 43–62.

730 Rose SM: A hierarchy of self-limiting reactions as the basis of cellular differentiation and growth control. Am Nat 1952;86:337–354.

731 Rose SM: Cellular interaction during differentiation. Biol Rev 1957;32:351–382.

732 Rose SM: Differentaiton during regeneration caused by migration of repressors in bioelectric fields. Am Zool 1970;10:91–99.

733 Ross J, Müller SC, Vidal C: Chemical waves. Science 1988;240:460–465.

734 Rossing TD: The physics of kettledrums. Sci Am 1982;247:172–178.

735 Rossing TD: The acoustics of bells. Am Sci 1984;72:440–447.

736 Roth VL: On homology. Biol J Linn Soc 1984;22:13–29.

737 Roth VL: The biological basis of homology; in Humphries CJ (ed): Ontogeny and Systematics. New York, Columbia University Press, 1988, pp 1–26.

738 Roux W: Der Kampf der Theile im Organismus. Ein Beitrag zur Vervollständigung der mechanischen Zweckmässigkeitslehre. Leipzig, Wilhelm Engelmann, 1881.

739 Roux W: The problems, methods, and scope of developmental mechanics. Biol Lects Marine Biol Lab Woods Hole 1895;(1894 Summer Session):149–190.

740 Rozenberg G, Salomaa A (eds): The Book of L. Berlin, Springer, 1986.

741 Rubin GM: Development of the *Drosophila* retina: Inductive events studied at single cell resolution. Cell 1989;57:519–520.

742 Rudensky AY, Rath S, Preston-Hurlburt P, Murphy DB, Janeway CA Jr: On the complexity of self. Nature 1991;353:660–662.

743 Ruiz-Gómez M, Modolell J: Deletion analysis of the *achaete-scute* locus of *Drosophila melanogaster*. Genes Dev 1987;1:1238–1246.

744 Runnström J: Über Selbstdifferenzierung und Induktion bei dem Seeigelkeim. Roux Arch Entwicklmech Org 1929;117:123–145.

745 Rushlow C, Levine M: Combinatorial expression of a *ftz-zen* fusion promoter suggests the occurrence of *cis* interactions between genes of the ANT-C. EMBO J 1988; 11:3479–3485.

746 Rushlow CA, Hogan A, Pinchin SM, Howe KM, Lardelli M, Ish-Horowicz D: The *Drosophila hairy* protein acts in both segmentation and bristle patterning and shows homology to N-*myc*. EMBO J 1989;8:3095–3103.

747 Russell MA: Positional information in imaginal discs: A Cartesian coordinate model; in Antonelli PL (ed): Mathematical Essays on Growth and the Emergence of Form. Edmonton, University of Alberta Press, 1985, pp 169–183.

748 Russell MA: Positional information in insect segments. Dev Biol 1985;108:269–283.

749 Russell MA, Girton JR, Morgan K: Pattern formation in a ts-cell-lethal mutant of *Drosophila:* The range of phenotypes induced by larval heat treatments. Roux Arch 1977;183:41–59.

750 Ruvkun G, Wightman B, Bürglin T, Arasu P: Dominant gain-of-function mutations that lead to misregulation of the *C. elegans* heterochronic gene *lin-14,* and the evolutionary implications of dominant mutations in pattern-formation genes. Development 1991;(suppl 1):47–54.

751 Sachs T: The development of spacing patterns in the leaf epidermis; in Subtelny S, Sussex IM (eds): The Clonal Basis of Development. Symp Soc Dev Biol, vol 36. New York, Academic Press, 1978, pp 161–183.

752 Sachs T: Patterned differentiation in plants. Differentiation 1978;11:65–73.

753 Sachs T: Axiality and polarity in vascular plants; in Barlow PW, Carr DJ (eds): Positional Controls in Plant Development. Cambridge, Cambridge University Press, 1984, pp 193–224.

754 Sachs T: Controls of cell patterns in plants; in Malacinski GM, Bryant SV (eds): Pattern Formation: A Primer in Developmental Biology. New York, Macmillan, 1984, pp 367–391.

755 Sander K: Formation of the basic body pattern in insect embryogenesis. Adv Insect Physiol 1976;12:125–238.

756 Sander K: The role of genes in ontogenesis – evolving concepts from 1883 to 1983 as perceived by an insect embryologist; in Horder TJ, Witkowski JA, Wylie CC (eds): A History of Embryology. Symp Br Soc Dev Biol, vol 8. Cambridge, Cambridge University Press, 1985, pp 363–395.

757 Sander K: Studies in insect segmentation: From teratology to phenogenetics. Development 1988;104(suppl):112–121.

758 Santamaria P: Analysis of haploid mosaics in *Drosophila.* Dev Biol 1983;96:285–295.

759 Sato T, Denell RE: Homoeosis in *Drosophila:* Anterior and posterior transformations of Polycomb lethal embryos. Dev Biol 1985;110:53–64.

760 Satoh N: Timing mechanisms in early embryonic development. Differentiation 1982;22:156–163.

761 Satoh N: Cell division cycles as the basis for timing mechanisms in early embryonic development of animals; in Edmunds LN Jr (ed): Cell Cycle Clocks. New York, Dekker, 1984, pp 527–538.

762 Satoh N: Recent advances in our understanding of the temporal control of early embryonic development in amphibians. J Embryol Exp Morphol 1985;89(suppl): 257–270.

763 Saunders JW Jr, Fallon JF: Cell death in morphogenesis; in Locke M (ed): Major Problems in Developmental Biology. Symp Soc Dev Biol, vol 25. New York, Academic Press, 1966, pp 289–314.

764 Saunders PT: The epigenetic landscape and evolution. Biol J Linn Soc 1990;39: 125–134.

765 Saunders PT, Ho MW: Primary and secondary waves in prepattern formation. J Theor Biol 1985;114:491–504.

766 Saunders PT, Kubal C: Bifurcations and the epigenetic landscape; in Goodwin B, Saunders P (eds): Theoretical Biology: Epigenetic and Evolutionary Order from Complex Systems. Edinburgh, Edinburgh University Press, 1989, pp 16–30.

767 Schlaggar BL, O'Leary DDM: Potential of visual cortex to develop an array of functional units unique to somatosensory cortex. Science 1991;252:1556–1560.

768 Schoute JC: Beiträge zur Blattstellungslehre. Réc Trav Bot Néerl 1913;10:153–235.

769 Schubiger G: Anlageplan, Determinationszustand und Transdeterminationsleistungen der männlichen Vorderbeinscheibe von *Drosophila melanogaster.* Roux Arch EntwicklMech Org 1968;160:9–40.

770 Schwalm FE: Insect Morphogenesis. Monogr Dev Biol, vol 20. Basel, Karger, 1988.

771 Schwartz LM: The role of cell death genes during development. BioEssays 1991;13: 389–395.

772 Schwartz RH: Acquisition of immunologic self-tolerance. Cell 1989;57:1073–1081.

773 Schwarz-Sommer Z, Huijser P, Nacken W, Saedler H, Sommer H: Genetic control of flower development by homeotic genes in *Antirrhinum majus.* Science 1990;250: 931–936.

774 Scott MP, Carroll SB: The segmentation and homeotic gene network in early *Drosophila* development. Cell 1987;51:689–698.

775 Sengel P: Feather pattern development; in Porter R, Rivers J (eds): Cell Patterning. Ciba Fd Symp, vol 29. New York, Elsevier, 1975, pp 51–70.

776 Sengel P: Morphogenesis of Skin. Cambridge, Cambridge University Press, 1976.

777 Seul M, Monar LR, O'Gorman L, Wolfe R: Morphology and local structure in labyrinthine stripe domain phase. Science 1991;254:1616–1618.

778 Shankland M: Positional determination of supernumerary blast cell death in the leech embryo. Nature 1984;307:541–543.

779 Shankland M: Leech segmentation: Cell lineage and the formation of complex body patterns. Dev Biol 1991;144:221–231.

780 Shatz CJ: Impulse activity and the patterning of connections during CNS development. Neuron 1990;5:745–756.

781 Shatz CJ, Kliot M: Prenatal misrouting of the retinogeniculate pathway in Siamese cats. Nature 1982;300:525–529.

782 Shatz CJ, LeVay S: Siamese cat: Altered connections of visual cortex. Science 1979; 204:328–330.

783 Shubin NH: A morphogenetic approach to the origin and basic organization of the tetrapod limb. Evol Biol 1986;20:319–387.

784 Sibatani A: Wing homoeosis in Lepidoptera: A survey. Dev Biol 1980;79:1–18.

785 Sibatani A: The Polar Co-ordinate Model for pattern regulation in epimorphic fields: A critical appraisal. J Theor Biol 1981;93:433–489.

786 Sibatani A: A plausible molecular interpretation of the polar coordinate model with a centripetal degeneracy of distinctive field values. J Theor Biol 1983;103:421–428.

787 Simcox AA, Hersperger E, Shearn A, Whittle JRS, Cohen SM: Establishment of imaginal discs and histoblast nests in *Drosophila.* Mech Dev 1991;34:11–20.

788 Simpson P: Lateral inhibition and the development of the sensory bristles of the adult peripheral nervous system of *Drosophila.* Development 1990;109:509–519.

789 Simpson P, Carteret C: Proneural clusters: equivalence groups in the epithelium of *Drosophila.* Development 1990;110:927–932.

790 Simpson P, Schneiderman HA: Isolation of temperature sensitive mutations blocking clone development in *Drosophila melanogaster*, and the effects of a temperature sensitive cell lethal mutation on pattern formation in imaginal discs. Roux Arch 1975;178:247–275.

791 Singer M, Berg P: Genes and Genomes. A Changing Perspective. Mill Valley, Science Books, 1990.

792 Singer SJ, Nicolson GL: The fluid mosaic model of the structure of cell membranes. Science 1972;175:720–731.

793 Slack JMW: A serial threshold theory of regeneration. J Theor Biol 1980;82:105–140.

794 Slack JMW: Regeneration and the second anatomy of animals; in Subtelny S, Green PB (eds): Developmental Order: Its Origin and Regulation. Symp Soc Dev Biol, vol 40. New York, Liss, 1982, pp 423–436.

795 Slack JMW: From Egg to Embryo: Determinative Events in Early Development. Cambridge, Cambridge University Press, 1983.

796 Slack JMW: The early amphibian embryo. A hierarchy of developmental decisions; in Malacinski GM, Bryant SV (eds): Pattern Formation: A Primer in Developmental Biology. New York, Macmillan, 1984, pp 457–480.

797 Slack JMW: Homoeotic transformations in man: Implications for the mechanism of embryonic development and for the organization of epithelia. J Theor Biol 1985; 114:463–490.

798 Slack JMW: Morphogenetic gradients – past and present. Trends Biochem Sci 1987; 12:200–204.

799 Small S, Kraut R, Hoey T, Warrior R, Levine M: Transcriptional regulation of a pair-rule stripe in *Drosophila.* Genes Dev 1991;5:827–839.

800 Smith WC, Harland RM: Injected Xwnt-8 RNA acts early in *Xenopus* embryos to promote formation of a vegetal dorsalizing center. Cell 1991;67:753–765.

801 Snow MHL, Tam PPL: Timing in embryological development. Nature 1980;286: 107

802 Sokol S, Christian JL, Moon RT, Melton DA: Injected Wnt RNA induces a complete body axis in *Xenopus* embryos. Cell 1991;67:741–752.

803 Sondhi KC: The biological foundations of animal patterns. Q Rev Biol 1963;38: 289–327.

804 Sonneborn TM: Determination, development, and inheritance of the structure of the cell cortex; in Padykula HA (ed): Control Mechanisms in the Expression of Cellular Phenotypes. Symp Int Soc Cell Biol, vol 9. New York, Academic Press, 1970, pp 1–13.

805 Spearman RIC: The Integument: A Textbook of Skin Biology. Biological Structure and Function, vol 3. Cambridge, Cambridge University Press, 1973.

806 Speksnijder JE, Sardet C, Jaffe LF: Periodic calcium waves cross ascidian eggs after fertilization. Dev Biol 1990;142:246–249.

807 Spemann H: Embryonic Development and Induction. New Haven, Yale University Press, 1938.

808 Sperry RW: Chemoaffinity in the orderly growth of nerve fiber patterns and connections. Proc Natl Acad Sci USA 1963;50:703–710.

809 Spiegel FW, Cox EC: A one-dimensional pattern in the cellular slime mould *Polysphondylium pallidum.* Nature 1980;286:806–807.

810 Spiegelman S: Physiological competition as a regulatory mechanism in morphogenesis. Q Rev Biol 1945;20:121–146.

811 Spitzer NC, Lamborghini JE: Programs of early neuronal development; in Cowman WM (ed): Studies in Developmental Neurobiology: Essays in Honor of Viktor Hamburger. New York, Oxford University Press, 1981, pp 261–287.

812 Sprague GF Jr: Combinatorial associations of regulatory proteins and the control of cell type in yeast. Adv Genet 1990;27:33–62.

813 Staddon BW, Edmunds MG: Gland regional and spatial patterns in the abdominal sternites of some pentatomoid Heteroptera. Ann Soc Ent Fr (NS) 1991;27:189–203.

814 Stanier RY, Cohen-Bazire G: Phototrophic prokaryotes: The cyanobacteria. Ann Rev Microbiol 1977;31:225–274.

815 Stanojevic D, Small S, Levine M: Regulation of a segmentation stripe by overlapping activators and repressors in the *Drosophila* embryo. Science 1991;254:1385–1387.

816 Stebbins GL, Jain SK: Developmental studies of cell differentiation in the epidermis of monocotyledons. I. *Allium, Rhoeo* and *Commelina.* Dev Biol 1960;2:409–426.

817 Stebbins GL, Shah SS: Developmental studies of cell differentiation in the epidermis of monocotyledons. II. Cytological features of stomatal development in the Gramineae. Dev Biol 1960;2:477–500.

818 Steeves TA, Sussex IM: Patterns in Plant Development, ed 2. Cambridge, Cambridge University Press, 1989.

819 Steinberg MS: Reconstruction of tissues by dissociated cells. Science 1963;141:401–408.

820 Steinberg MS: Does differential adhesion govern self-assembly processes in histogenesis? Equilibrium configurations and the emergence of a hierarchy among populations of embryonic cells. J Exp Zool 1970;173:395–434.

821 Steinberg MS: The adhesive specification of tissue self-organization; in Connelly TG, Brinkley LL, Carlson BM (eds): Morphogenesis and Pattern Formation. New York, Raven Press, 1981, pp 179–203.

822 Steinberg MS, Poole TJ: Strategies for specifying form and pattern: Adhesion-guided multicellular assembly. Phil Trans R Soc Lond [B] 1981;295:451–460.

823 Steinberg MS, Poole TJ: Cellular adhesive differentials as determinants of morphogenetic movements and organ segregation; in Subtelny S, Green PB (eds): Developmental Order: Its Origin and Regulation. Symp Soc Dev Biol, vol 40. New York, Liss, 1982, pp 351–378.

824 Steiner E: Establishment of compartments in the developing leg imaginal discs of *Drosophila melanogaster.* Roux Arch 1976;180:9–30.

825 Stent GS: Paradoxes of Progress. San Francisco, Freeman, 1978.

826 Stent GS: Strength and weakness of the genetic approach to the development of the nervous system. Ann Rev Neurosci 1981;4:163–194.

827 Stent GS: From probability to molecular biology. Cell 1984;36:567–571.

828 Stent GS: Hermeneutics and the analysis of complex biological systems; in Depew DJ, Weber BH (eds): Evolution at a Crossroads: The New Biology and the New Philosophy of Science. Cambridge, MIT Press, 1985, pp 209–225.

829 Stent GS: The role of cell lineage in development. Phil Trans R Soc Lond [B] 1985;312:3–19.

830 Stent GS: Thinking in one dimension: The impact of molecular biology on development. Cell 1985;40:1–2.

831 Stern C: Genes and developmental patterns. Caryologia 1954;(suppl)6:355–369.

832 Stern C: Two or three bristles. Am Scient 1954;42:213–247.

833 Stern C: The genetic control of developmental competence and morphogenetic tissue interactions in genetic mosaics. Roux Arch Dev Biol 1956;149:1–25.

834 Stern C: Genetic mechanisms in the localized initiation of differentiation. Cold Spring Harb Symp Quant Biol 1956;21:375–382.

835 Stern C: Developmental genetics of pattern; in: Genetic Mosaics and Other Essays. Cambridge, Harvard University Press, 1968, pp 130–173.

836 Stern CD, Fraser SE, Keynes RJ, Primmett DRN: A cell lineage analysis of segmentation in the chick embryo. Development 1988;104(suppl):231–244.

837 Stern CD, Goodwin BC: Waves and periodic events during primitive streak formation in the chick. J Embryol Exp Morphol 1977;41:15–22.

838 Stern MJ, Horvitz HR: A normally attractive cell interaction is repulsive in two *C. elegans* mesodermal cell migration mutants. Development 1991;113:797–803.

839 Sternberg PW: Lateral inhibition during vulval induction in *Caenorhabditis elegans.* Nature 1988;335:551–554.

840 Sternberg PW: Genetic control of cell type and pattern formation in *Caenorhabditis elegans.* Adv Genet 1990;27:63–116.

841 Sternberg PW, Horvitz HR: Gonadal cell lineages of the nematode *Panagrellus redivivus* and implications for evolution by the modification of cell lineage. Dev Biol 1981;88:147–166.

842 Sternberg PW, Horvitz HR: Postembryonic nongonadal cell lineages of the nematode *Panagrellus redivivus:* Description and comparison with those of *Caenorhabditis elegans.* Dev Biol 1982;93:181–205.

843 Sternberg PW, Horvitz HR: The genetic control of cell lineage during nematode development. Ann Rev Genet 1984;18:489–524.

844 Sternberg PW, Horvitz HR: *lin*-17 mutations of *Caenorhabditis elegans* disrupt certain asymmetric cell divisions. Dev Biol 1988;130:67–73.

845 Sternberg PW, Horvitz HR: The combined action of two intercellular signaling pathways specifies three cell fates during vulval induction in *C. elegans.* Cell 1989;58:679–693.

846 Sternfeld J, David CN: Cell sorting during pattern formation in *Dictyostelium.* Differentiation 1981;20:10–21.

847 Stewart WDP, Haystead A, Pearson HW: Nitrogenase activity in heterocysts of blue-green algae. Nature 1969;224:226–228.

848 Stock GB, Bryant SV: Studies of digit regeneration and their implications for theories of development and evolution of vertebrate limbs. J Exp Zool 1981;216:423–433.

849 Storey KG: Cell lineage and pattern formation in the earthworm embryo. Development 1989;107:519–531.

850 Struhl G: A gene product required for correct initiation of segmental determination in *Drosophila.* Nature 1981;293:36–41.

851 Struhl G: Spineless-aristapedia: A homeotic gene that does not control the development of specific compartments in *Drosophila.* Genetics 1982;102:737–749.

852 Struhl G: Role of the *esc*+ gene product in ensuring the selective expression of segment-specific homeotic genes in *Drosophila.* J Embryol Exp Morphol 1983;76:297–331.

853 Struhl G: Near-reciprocal phenotypes caused by inactivation or indiscriminate expression of the *Drosophila* segmentation gene *ftz.* Nature 1985;318:677–680.

854 Struhl G: Morphogen gradients and the control of body pattern in insect embryos; in Evered D, Marsh J (eds): Cellular Basis of Morphogenesis. Ciba Found Symp, vol 144. New York, Wiley, 1989, pp 65–91.

855 Struhl G, Brower D: Early role of the *esc+* gene product in the determination of segments in *Drosophila.* Cell 1982;31:285–292.

856 Struhl G, Struhl K, Macdonald PM: The gradient morphogen bicoid is a concentration-dependent transcriptional activator. Cell 1989;57:1259–1273.

857 Stuart JJ, Brown SJ, Beeman RW, Denell RE: A deficiency of the homeotic complex of the beetle *Tribolium.* Nature 1991;350:72–74.

858 Stumpf HF: Mechanism by which cells estimate their location within the body. Nature 1966;212:430–431.

859 Sturtevant AH: The use of mosaics in the study of the developmental effects of genes. Proc 6th Int Congr Genet 1932;1:304–307.

860 Subtelny S, Konigsberg IR (eds): Determinants of Spatial Organization. Symp Soc Dev Biol, vol 37. New York, Academic Press, 1979.

861 Subtelny S, Sussex IM (eds): The Clonal Basis of Development. Symp Soc Dev Biol, vol 36. New York, Academic Press, 1978.

862 Süffert F: Die Geschichte der Bildungszellen im Puppenflügelepithel bei einem Tagschmetterling. Biol Zentralbl 1937;57:615–628.

863 Sullivan W: Independence of *fushi tarazu* expression with respect to cellular density in *Drosophila* embryos. Nature 1987;327:164–167.

864 Sulston JE, Albertson DG, Thomson JN: The *Caenorhabditis elegans* male: Postembryonic development of nongonadal structures. Dev Biol 1980;78:542–576.

865 Sulston JE, Schierenberg E, White JG, Thomson JN: The embryonic cell lineage of the nematode *Caenorhabditis elegans.* Dev Biol 1983;100:64–119.

866 Summerbell D, Honig LS: The control of pattern across the antero-posterior axis of the chick limb bud by a unique signalling region. Am Zool 1982;22:105–116.

867 Summerbell D, Lewis JH: Time, place and positional value in the chick limb-bud. J Embryol Exp Morphol 1975;33:621–643.

868 Summerbell D, Lewis JH, Wolpert L: Positional information in chick limb morphogenesis. Nature 1973;244:492–496.

869 Sumper M: Control of differentiation in *Volvox carteri.* FEBS Lett 1979;107:241–246.

870 Sumper M: Pattern formation during embryogenesis of the multicellular organism *Volvox;* in Malacinski, Bryant SV (eds): Pattern Formation: A Primer in Developmental Biology. New York, Macmillan, 1984, pp 197–212.

871 Sussex IM: Developmental programming of the shoot meristem. Cell 1989;56:225–229.

872 Swindale NV: A model for the formation of ocular dominance stripes. Proc R Soc Lond [B] 1980;208:243–264.

873 Tabin CJ: Isolation of potential vertebrate limb-identity genes. Development 1989;105:813–820.

874 Takeuchi I: Cell sorting and pattern formation in *Dictyostelium discoideum;* in Gerhart JC (ed): Cell-Cell Interactions in Early Development. Symp Soc Dev Biol, vol 49. New York, Wiley, 1991, pp 249–259.

875 Taylor SS: Protein kinases: A diverse family of related proteins. BioEssays 1987;7:24–29.

876 Temple S, Raff MC: Clonal analysis of oligodendrocyte development in culture: evidence for a developmental clock that counts cell divisions. Cell 1986;44:773–779.

877   Tennant NW: Reductionism and holism in biology; in Horder TJ, Witkowski JA, Wylie CC (eds): A History of Embryology. Symp Br Soc Dev Biol, vol 8. Cambridge, Cambridge University Press, 1985, pp 407–433.

878   Tesler LG: Programming languages. Sci Am 1984;251:70–78.

879   Teuber ML: Sources of ambiguity in the prints of Maurits C. Escher. Sci Am 1974; 231:90–104.

880   Thiebaud P, Goodstein M, Calzone FJ, Thézé N, Britten RJ, Davidson EH: Intersecting batteries of differentially expressed genes in the early sea urchin embryo. Genes Dev 1990;4:1999–2010.

881   Thom R: Gradients in biology, in mathematics, and simultaneous optimization; in: Lectures on Mathematics in the Life Sciences, vol 6. Providence, American Mathematics Society, 1974, pp 1–13.

882   Thompson DW: On Growth and Form, ed 2. Cambridge, Cambridge University Press, 1942.

883   Thornley JHM: Phyllotaxis. I. A mechanistic model. Ann Bot 1975;39:491–507.

884   Thornley JHM: Phyllotaxis. II. A description in terms of intersecting logarithmic spirals. Ann Bot 1975;39:509–524.

885   Thummel CS: Puffs and gene regulation – molecular insights into the *Drosophila* ecdysone regulatory hierarchy. BioEssays 1990;12:561–568.

886   Toffoli T, Margolus N: Cellular Automata Machines: A New Environment for Modeling. Cambridge, MIT Press, 1987.

887   Tokunaga C: Autonomy or nonautonomy of gene effects in mosaics. Proc Natl Acad Sci USA 1972;69:3283–3286.

888   Tokunaga C: Genetic mosaic studies of pattern formation in *Drosophila melanogaster,* with special reference to the prepattern hypothesis; in Gehring WJ (ed): Genetic Mosaics and Cell Differentiation. New York, Springer, 1978, pp 157–204.

889   Tokunaa C, Stern C: The developmental autonomy of extra sex combs in *Drosophila melanogaster.* Dev Biol 1965;11:50–81.

890   Tomlinson A: Cellular interactions in the developing *Drosophila* eye. Development 1988;104:183–193.

891   Tomlinson A, Ready DF: *Sevenless:* A cell-specific homeotic mutation of the *Drosophila* eye. Science 1986;231:400–402.

892   Tomlinson A, Ready DF: Cell fate in the *Drosophila* ommatidium. Dev Biol 1987; 123:264–275.

893   Tomlinson A, Ready DF: Neuronal differentiation in the *Drosophila* ommatidium. Dev Biol 1987;120:366–376.

894   Tompkins R, Szaro B, Reinschmidt D, Kaye C, Ide C: Effects of alterations of cell size and number on the structure and function of the *Xenopus laevis* nervous system; in Lauder JM, Nelson PG (eds): Gene Expression and Cell-Cell Interactions in the Developing Nervous System. New York, Plenum Press, 1984, pp 135–146.

895   Torres M, Sánchez L: The *scute* (T4) gene acts as a numerator element of the X:A signal that determines the state of activity of *Sex-lethal* in *Drosophila.* EMBO J 1989;8:3079–3086.

896   Toussaint N, French V: The formation of pattern on the wing of the moth, *Ephestia kühniella.* Development 1988;103:707–718.

897   Townes PL, Holtfreter J: Directed movements and selective adhesion of embryonic amphibian cells. J Exp Zool 1955;128:53–120.

898 Trinkaus JP: Morphogenetic cell movements; in Locke M (ed): Major Problems in Developmental Biology. Symp Soc Dev Biol, vol 25. New York, Academic Press, 1966, pp 125–176.

899 Trinkaus JP: Cells into Organs: The Forces That Shape the Embryo, ed 2. Englewood Cliffs, Prentice-Hall, 1984.

900 Trinkaus JP: Further thoughts on directional cell movement during morphogenesis. J Neurosci Res 1985;13:1–19.

901 Trinkaus JP: Directional cell movement during early development of the teleost *Blennius pholis.* I. Formation of epithelial cell clusters and their pattern and mechanism of movement. J Exp Zool 1988;245:157–186.

902 Trisler D, Collins F: Corresponding spatial gradients of TOP molecules in the developing retina and optic tectum. Science 1987;237:1208–1209.

903 Truman JW, Bate M: Spatial and temporal patterns of neurogenesis in the central nervous system of *Drosophila melanogaster.* Dev Biol 1988;125:145–157.

904 Tucker RP, Erickson CA: The control of pigment cell pattern formation in the California newt, *Taricha torosa.* J Embryol Exp Morphol 1986;97:141–168.

905 Tucker RP, Erickson CA: Pigment cell pattern formation in *Taricha torosa:* The role of the extracellular matrix in controlling pigment cell migration and differentiation. Dev Biol 1986;118:268–285.

906 Turing AM: On computable numbers, with an application to the Entscheidungs problem. Proc Lond Math Soc [2] 1937;42:230–265.

907 Turing AM: The chemical basis of morphogenesis. Phil Trans R Soc Lond [B] 1952; 237:37–72.

908 Twitty VC: Chromatophore migration as a response to mutual influences of the developing pigment cells. J Exp Zool 1944;95:259–290.

909 Twitty VC: The developmental analysis of specific pigment patterns. J Exp Zool 1945;100:141–178.

910 Twitty VC, Bodenstein D: The effect of temporal and regional differentials on the development of grafted chromatophores. J Exp Zool 1944;95:213–232.

911 Twitty VC, Niu MC: Causal analysis of chromatophore migration. J Exp Zool 1948; 108:405–437.

912 Twitty VC, Niu MC: The motivation of cell migration, studied by isolation of embryonic pigment cells singly and in small groups in vitro. J Exp Zool 1954;125: 541–573.

913 Tyagi VVS: The heterocysts of blue-green algae (Myxophyceae). Biol Rev 1975;50: 247–284.

914 Tyson JJ: Cyclic-AMP waves in *Dictyostelium:* specific models and general theories; in Goldbeter A (ed): Cell to Cell Signalling: From Experiments to Theoretical Models. New York, Academic Press, 1989, pp 521–537.

915 Tyson JJ, Murray JD: Cyclic AMP waves during aggregation of *Dictyostelium* amoebae. Development 1989;106:421–426.

916 Uemura T, Shepherd S, Ackerman L, Jan LY, Jan YN: *numb,* a gene required in determination of cell fate during sensory organ formation in *Drosophila* embryos. Cell 1989;58:349–360.

917 Ursprung H: The formation of patterns in development; in Locke M (ed): Major Problems in Developmental Biology. Symp Soc Dev Biol, vol 25. New York, Academic Press, 1966, pp 177–216.

918 Van Doren M, Ellis HM, Posakony JW: The *Drosophila extramacrochaetae* protein antagonizes sequence-specific DNA binding by *daughterless/achaete-scute* protein complexes. Development 1991;113:245–255.

919 van Oss CJ: An explanation of the Liesegang phenomenon. Science 1959;129:1365–1366.

920 Van Valen LM: Homology and causes, J Morphol 1982;173:305–312.

921 Varner JE (ed): Self-Assembling Architecture. Symp Soc Dev Biol, vol 46. New York, Liss, 1988.

922 Vollrath F: Altered geometry of webs in spiders with regenerated legs. Nature 1987; 328:247–248.

923 von Boehmer H, Kisielow P: How the immune system learns about self. Sci Am 1991;265:74–81.

924 von Neumann J: Theory of Self-Reproducing Automata. Urbana, University of Illinois Press, 1966.

925 Waddington CH: Organizers and Genes. Cambridge, Cambridge University Press, 1940.

926 Waddington CH: The Strategy of the Genes: A Discussion of Some Aspects of Theoretical Biology. London, George Allen & Unwin, 1957.

927 Waddington CH: New Patterns in Genetics and Development. New York, Columbia University Press, 1962.

928 Waddington CH: Autogenous cellular periodicities as (a) temporal templates and (b) the basis of 'morphogenetic fields'. J Theor Biol 1965;8:367–369.

929 Waddington CH: Fields and gradients; in Locke M (ed): Major Problems in Developmental Biology. Symp Soc Dev Biol, vol 25. New York, Academic Press, 1966, pp 105–124.

930 Waddington CH: The theory of evolution today; in Koestler A, Smythies JR (eds): Beyond Reductionism. New Perspectives in the Life Sciences. New York, MacMillan, 1969, pp 357–395.

931 Waddington CH: The morphogenesis of patterns in *Drosophila;* in Counce SJ, Waddington CH (eds): Developmental Systems: Insects, vol 2. New York, Academic Press, 1973, pp 499–535.

932 Wagner GP: The biological homology concept. Ann Rev Ecol Syst 1989;20:51–69.

933 Wagner GP: The origin of morphological characters and the biological basis of homology. Evolution 1989;43:1157–1171.

934 Walker TM: Introduction to Computer Science. Boston, Allyn-Bacon, 1972.

935 Ward PD, Chamberlain J: Radiographic observation of chamber formation in *Nautilus pompilius*. Nature 1983;304:57–59.

936 Wardlaw CW: Experiments on organogenesis in ferns. Growth 1949;13(suppl):93–131.

937 Warrior R, Levine M: Dose-dependent regulation of pair-rule stripes by gap proteins and the initiation of segment polarity. Development 1990;110:759–767.

938 Way JC, Chalfie M: *mec-3*, a homeobox-containing gene that specifies differentiation of the touch receptor neurons in *C. elegans.* Cell 1988;54:5–16.

939 Way JC, Chalfie M: The *mec-3* gene of *Caenorhabditis elegans* requires its own product for maintained expression and is expressed in three neuronal cell types. Genes Dev 1989;3:1823–1833.

940 Way JC, Wang L, Run J-Q, Wang A: The *mec-3* gene contains *cis*-acting *elements* mediating positive and negative regulation in cells produced by asymmetric cell division in *Caenorhabditis elegans.* Genes Dev 1991;5:2199–2211.

941 Weintraub H: Assembly and propagation of repressed and derepressed chromosomal states. Cell 1985;42:705–711.

942 Weintraub H, Davis R, Tapscott S, Thayer M, Krause M, Benezra R, Blackwell TK, Turner D, Rupp R, Hollenberg S, Zhuang Y, Lassar A: The *myoD* gene family: Nodal point during specification of the muscle cell lineage. Science 1991;251:761–766.

943 Weintraub H, Dwarki VJ, Verma I, Davis R, Hollenberg S, Snider L, Lassar A, Tapscott SJ: Muscle-specific transcriptional activation by MyoD. Genes Dev 1991; 5:1377–1386.

944 Weiss P: Functional adaptation and the rôle of ground substances in development. Am Nat 1933;67:322–340.

945 Weiss P: Principles of Development. New York, Holt, 1939.

946 Weiss P: The problem of cell individuality in development. Am Nat 1940;74:34–46.

947 Weiss P. The problem of specificity in growth and development. Yale J Biol Med 1947;19:235–278.

948 Weiss P: Some introductory remarks on the cellular basis of differentiation. J Embryol Exp Morphol 1953;1:181–211.

949 Weiss P: Beauty and the beast: Life and the rule of order. Sci Monthly 1955;81: 286–299.

950 Weiss P: Cellular dynamics. Rev Mod Phys 1959;31:11–50.

951 Weiss P: Interactions between cells. Rev Mod Phys 1959;31:449–454.

952 Weiss P: From cell to molecule; in Allen JM (ed): The Molecular Control of Cellular Activity. New York, McGraw-Hill, 1961, pp 1–72.

953 Weiss P: Structure as the coordinating principle in the life of the cell. Proc Welch Found Conf Chem Res 1961;5:5–31.

954 Weiss P: The cell as unit. J Theor Biol 1963;5:389–397.

955 Weiss P: The living system: determinism stratified; in Koestler A, Smythies JR (eds): Beyond Reductionism: New Perspectives in the Life Sciences. New York, MacMillan, 1969, pp 3–55.

956 Wessells NK: A catalogue of processes responsible for metazoan morphogenesis; in Bonner JT (ed): Evolution and Development. New York, Springer, 1982, pp 115–154.

957 Whimster IW: The focal differentiation of pigment cells. J Exp Zool 1979;208:153–159.

958 White RH: Analysis of the development of the compound eye in the mosquito. *Aedes aegypti.* J Exp Zool 1961;148:223–239.

959 Whitten JM: Comparative anatomy of the tracheal system. Ann Rev Ent 1972;17: 373–402.

960 Whitten JM: Some observations on cellular organization and pattern in flies; in Hepburn HR (ed): The Insect Integument. New York, Elsevier, 1976, pp 277–297.

961 Whittle JRS: Litany and creed in the genetic analysis of development; in Goodwin BC, Holder N, Wylie CC (eds): Development and Evolution. Symp Br Soc Dev Biol, vol 6. Cambridge, Camridge University Press, 1983, pp 59–74.

962 Whittle JRS: The developmental programme – concept or muddle. BioEssays 1986; 5:91–92.

963 Wigglesworth VB: Local and general factors in the development of 'pattern' in *Rhodnius prolixus* (Hemiptera). J Exp Zool 1940;17:180–200.

964 Wigglesworth VB: The control of pattern as seen in the integument of an insect. Bio-Essays 1988;9:23–27.

965 Wikler KC, Rakic P: Relation of an array of early-differentiating cones to the photoreceptor mosaic in the primate retina. Nature 1991;351:397–400.

966 Wilby OK, Ede DA: A model generating the pattern of cartilage skeletal elements in the embryonic chick limb. J Theor Biol 1975;52:199–217.

967 Wilcox M, Mitchison GJ, Smith RJ: Pattern formation in the blue-green alga, *Anabaena*. I. Basic mechanisms. J Cell Sci 1973;12:707–723.

968 Wilcox M, Mitchison GJ, Smith RJ: Pattern formation in the blue-green alga, *Anabaena*. II. Controlled proheterocyst regression. J Cell Sci 1973;13:637–649.

969 Wilcox M, Mitchison GJ, Smith RJ: Mutants of *Anabaena cylindrica* altered in heterocyst spacing. Arch Microbiol 1975;103:219–223.

970 Wilkins AS: The limits of molecular biology. BioEssays 1985;3:3.

971 Wilkins AS, Gubb D: Pattern formation in the embryo and imaginal discs of *Drosophila:* What are the links? Dev Biol 1991;145:1–12.

972 Williams GJA, Shivers RR, Caveney S: Active muscle migration during insect metamorphosis. Tiss Cell 1984;16:411–432.

973 Williams RW, Herrup K: The control of neuron number. Ann Rev Neurosci 1988; 11:423–453.

974 Williamson AR, Zitron IM, McMichael AJ: Clones of B lymphocytes: Their natural selection and expansion. Fed Proc 1976;35:2195–2201.

975 Wilson EB: The Cell in Development and Inheritance. New York, Macmillan, 1900.

976 Winfree AT: Rotating chemical reactions. Sci Am 1974;230:82–95.

977 Winfree AT: The Geometry of Biological Time. New York, Springer, 1980.

978 Winfree AT: A continuity principle for regeneration; in Malacinski GM, Bryant SV (eds): Pattern Formation: A Primer in Developmental Biology. New York, Macmillan, 1984, pp 103–124.

979 Winfree AT: The prehistory of the Belousov-Zhabotinsky oscillator. J Chem Ed 1984;61:661–663.

980 Winfree AT: Organizing centers for chemical waves in two and three dimensions; in Field RJ, Burger M (eds): Oscillations and Traveling Waves in Chemical Systems. New York, Wiley, 1985, pp 441–472.

981 Winfree AT: Crystals from dreams. Nature 1991;352:568–569.

982 Winfree AT, Winfree EM, Seifert H: Organizing centers in a cellular excitable medium. Physica 1985;17D:109–115.

983 Winklbauer R, Hausen P: Development of the lateral line system in *Xenopus laevis*. I. Normal development and cell movement in the supraorbital system. J Embryol Exp Morphol 1983;76:265–281.

984 Winklbauer R, Hausen P: Development of the lateral line system in *Xenopus laevis*. IV. Pattern formation in the supraorbital system. J Embryol Exp Morphol 1985;88: 193–207.

985 Witt PN, Reed CF: Spider-web building. Science 1965;149:1190–1197.

986 Wolff T, Ready DF: The beginning of pattern formation in the *Drosophila* compound eye: The morphogenetic furrow and the second mitotic wave. Development 1991;113:841–850.

987 Wolff T, Ready DF: Cell death in normal and rough eye mutants of *Drosophila*. Development 1991;113:825–839.

988 Wolfram S: Cellular automata as models of complexity. Nature 1984;311:419–424.

989 Wolfram S: Computer software in science and mathematics. Sci Am 1984;251:188–203.

990 Wolk CP: Physiological basis of the pattern of vegetative growth of a blue-green alga. Proc Natl Acad Sci USA 1967;57:1246–1251.

991 Wolk CP: Movement of carbon from vegetative cells to heterocysts in *Anabaena cylindrica*. J Bacteriol 1968;96:2138–2143.

992 Wolk CP: Differentiation and pattern formation in filamentous blue-green algae; in Gerhardt P, Costilow RN, Sadoff HL (eds): Spores VI. Washington, American Society of Microbiology, 1975, pp 85–96.

993 Wolk CP: Intercellular interactions and pattern formation in filamentous cyanobacteria; in Subtelny S, Konigsberg IR (eds): Determinants of Spatial Organization. Symp Soc Dev Biol, vol 37. New York, Academic Press, 1979, pp 247–266.

994 Wolk CP: Alternative models for the development of the pattern of spaced heterocysts in *Anabaena (Cyanophyta)*. Pl Syst Evol 1989;164:27–31.

995 Wolk CP, Quine MP: Formation of one-dimensional patterns by stochastic processes and by filamentous blue-green algae. Dev Biol 1975;46:370–382.

996 Wolpert L: The French Flag Problem: A contribution to the discussion on pattern development and regulation; in Waddington CH (ed): Towards a Theoretical Biology. I. Prolegomena. Chicago, Aldine, 1968, pp 125–133.

997 Wolpert L: Positional information and the spatial pattern of cellular differentiation. J Theor Biol 1969;25:1–47.

998 Wolpert L: Positional information and pattern formation. Curr Top Dev Biol 1971;6:183–224.

999 Wolpert L: Positional information and the development of pattern and form; in: Lectures on Mathematics in the Life Sciences, vol 6. Providence, American Mathematics Society, 1974, pp 27–41.

1000 Wolpert L: Cell position and cell lineage in pattern formation and regulation; in Lord B, Potten CS, Cole R (eds): Stem Cells and Tissue Homeostasis. Br Soc Cell Biol Symp, vol 2. Cambridge, Cambridge University Press, 1978, pp 29–47.

1001 Wolpert L: Positional information and pattern formation. Phil Trans R Soc Lond [B] 1981;295:441–450.

1002 Wolpert L: Pattern formation and change; in Bonner JT (ed): Evolution and Development. New York, Springer, 1982, pp 169–188.

1003 Wolpert L: Constancy and change in the development and evolution of pattern; in Goodwin BC, Holder N, Wylie CC (eds): Development and Evolution. Symp Br Soc Dev Biol, vol 6. Cambridge, Cambridge University Press, 1983, pp 47–57.

1004 Wolpert L: Molecular problems of positional information. BioEssays 1984;1:175–177.

1005 Wolpert L: Gradients, position and pattern: a history; in Horder TJ, Witkowski JA, Wylie CC (eds): A History of Embryology. Symp Br Soc Dev Biol, vol 8. New York, Cambridge University Press, 1985, pp 347–362.

1006 Wolpert L: Stem cells: A problem in asymmetry. J Cell Sci 1988;10(suppl):1–9.

1007 Wolpert L: Positional information and prepattern in the development of pattern; in Goldbeter A (ed): Cell to Cell Signalling: From Experiments to Theoretical Models. New York, Academic Press, 1989, pp 133–143.

1008 Wolpert L: Positional information revisited. Development 1989;(suppl):3–12.

1009 Wolpert L: Positional information. Semin Dev Biol 1991;2:77–82.

1010 Wolpert L, Lewis JH: Towards a theory of development. Fed Proc 1975;34:14–20.

1011 Wolpert L, Stein WD: Positional information and pattern formation; in Malacinski GM, Bryant SV (eds): Pattern Formation: A Primer in Developmental Biology. New York, Macmillan, 1984, pp 3–21.

1012 Wyman RJ: Sequential induction and a homeotic switch of cell fate. Trends Neurosci 1986;9:339–340.

1013 Yates FE: Systems analysis of hormone action: Principles and strategies; in Goldberger RF, Yamamoto KR (eds): Hormone Action. Biological Regulation and Development, vol 3A. New York, Plenum Press, 1982, pp 25–97.

1014 Yoshida A, Aoki K: Scale arrangement pattern in a lepidopteran wing. 1. Periodic cellular pattern in the pupal wing of *Pieris rapae*. Dev Growth Differ 1989;31:601–609.

1015 Yoshida A, Shinkawa T, Aoki K: Periodical arrangement of scales on lepidopteran (butterfly and moth) wings. Proc Jpn Acad [B] 1983;59:236–239.

1016 Young DA: On the diffusion theory of phyllotaxis. J Theor Biol 1978;71:421–432.

1017 Zackson SL, Steinberg MS: Cranial neural crest cells exhibit directed migration on the pronephric duct pathway: Further evidence for an in vivo adhesion gradient. Dev Biol 1986;117:342–353.

1018 Zackson SL, Steinberg MS: Chemotaxis or adhesion gradient? Pronephric duct elongation does not depend on distant sources of guidance information. Dev Biol 1987;124:418–422.

1019 Zahs KR: Influence of neural activity on the development and plasticity of the cat's visual cortex; in Landmesser LT (ed): The Assembly of the Nervous System. Symp Soc Dev Biol, vol 47. New York, Liss, 1989, pp 259–278.

1020 Zeeman EC: Primary and secondary waves in developmental biology; in: Lectures on Mathematics in the Life Sciences, vol 7. Providence, American Mathematics Society, 1974, pp 69–161.

1021 Zuckerkandl E: Programs of gene action and progressive evolution; in: Goodman M, Tashian RE, Tashian JH (eds): Molecular Anthropology: Genes and Proteins in the Evolutionary Ascent of the Primates. New York, Plenum Press, 1976, pp 387–447.